CLIMATE CATASTROPHE!

SCIENCE or SCIENCE FICTION?

BY ANDY MAY

COVER IMAGE CREDITS

The solar image is from NASA SpacePlace: https://spaceplace.nasa.gov/gallery-sun/en/
Public domain ultraviolet image of the Sun.

The Earth image is a true color satellite image and was created by Reto Stockli, Nazmi El
Saleous and Marit Jentoft-Nilsen of NASA GSFC. The site is:
https://earthobservatory.nasa.gov/IOTD/view.php?id=88 Public domain image.

American Freedom Publications LLC
www.americanfreedompublications.com
2638 E. Wildwood Road
Springfield, MO 65804
ISBN 978-1-64255-442-7   Hardback
ISBN 978-1-64255-443-4   Paperback
ISBN 978-1-64255-444-1   eBook
Cover Design Christopher M. Capages
www.capagescreative.com
Manuscript Editor Jordan Reilly Helterbrand

First Edition- May 1, 2018

Printed in the United States of America

DEDICATION

This book is dedicated to science. Scientists are skeptical, we ask: "Is that idea correct? How can I test it?" Then we resolve to gather and analyze data until we show it isn't or it might be. If we cannot disprove the idea, it survives. No true scientist "believes in science" because he knows science is a process, a process we use to uncover the truth. One cannot have faith in science, but one can believe in the scientific process or method. *Nullius in verba,* take nobody's word for it. This is the motto of the Royal Society; unfortunately, the Royal Society leadership has not taken this motto to heart (Carter, et al. 2015).

ACKNOWLEDGMENTS

I gratefully acknowledge the help of the gang at Wattsupwiththat.com, especially Anthony Watts. Without them, this book would never have been written.

I also acknowledge the infinite patience of my wife Aurelia as I worked on this book.

TABLE OF CONTENTS

DEDICATION V
ACKNOWLEDGMENTS VI
TABLE OF CONTENTS VII
FOREWORD XI
PREFACE XIII

CHAPTER 1 1

DO HUMANS HARM THE ENVIRONMENT? 1
CAN GLOBAL WARMING DESTROY THE EARTH? 2
THE OCEANS ARE KEY 3
ECONOMIC GROWTH, PROSPERITY, AND THE ENVIRONMENT 6
DISEASE AND HEALTH 8
LIFE EXPECTANCY 16
IS THE ENVIRONMENT DETERIORATING? 22
PROSPERITY 22
CHAPTER 1 CONCLUSIONS 25
CHAPTER 1 SUMMARY 29

CHAPTER 2 31

POPULATION GROWTH AND THE FOOD SUPPLY 31
POPULATION GROWTH 32
THE EFFECT OF CO_2 AND GLOBAL WARMING ON OUR FOOD SUPPLY 38
MALNOURISHED CHILDREN, WAR AND CLIMATE CHANGE 40
FOOD PRODUCTION 43
CHAPTER 2 CONCLUSIONS 55
CHAPTER 2 SUMMARY 56

CHAPTER 3 59

THE COST OF GLOBAL WARMING 59
WHAT IS THE ACTUAL COST OF GLOBAL WARMING? 60
MITIGATION AND ADAPTATION 63
THE TIME VALUE OF MONEY AND THE DISCOUNT RATE 64
THE COST OF REDUCING CO_2 EMISSIONS 67
THE PARIS CAGREEMENT 68
COSTS VERSUS BENEFITS 72
CHAPTER 3 CONCLUSIONS 76
CHAPTER 3 SUMMARY 79

CHAPTER 4 81

GLOBAL TEMPERATURES IN THE HOLOCENE 81
HOLOCENE CLIMATIC OVERVIEW 83
OCEAN TEMPERATURES AND THEIR SIGNIFICANCE 84
GREENLAND ICE CORES 90
GREENLAND VERSUS THE REST OF THE WORLD 92
EXAMINING THE MARCOTT, ET AL. PROXIES 95
PROXY SELECTION 98
SUMMARY OF THE RECONSTRUCTIONS 108
    ANTARCTIC (90S TO 60S) 108
    MID-LATITUDE SOUTHERN HEMISPHERE 110
    TROPICS.. 112
    NORTHERN HEMISPHERE MID-LATITUDES 116
    THE ARCTIC 118
THE GLOBAL RECONSTRUCTION 121
A WORD ABOUT ERROR.. 125
CHAPTER 4 CONCLUSIONS 127
CHAPTER 4 SUMMARY... 130

CHAPTER 5 133

EXTINCTIONS AND THE GULF STREAM 133
GLOBAL WARMING IS CAUSING A "GREAT EXTINCTION" EVENT............. 133
THE "GREAT" QUATERNARY MEGAFAUNA EXTINCTION 140
POLAR BEARS 147
GLOBAL WARMING WILL SHUT DOWN THE GULF STREAM AND CAUSE A
MINI-ICE AGE 150
CHAPTER 5 CONCLUSIONS 155
CHAPTER 5 SUMMARY... 157

CHAPTER 6 159

CLIMATE-RELATED DEATHS AND INSECURITY 159
THERE WILL BE MORE HEAT-RELATED DEATHS 159
INSECURITY 167
CHAPTER 6 CONCLUSIONS 170
CHAPTER 6 SUMMARY... 172

CHAPTER 7... **173**
GLOBAL WARMING AND EXTREME WEATHER.................................173
GLOBAL WARMING AND THE FREQUENCY AND SEVERITY OF NATURAL DISASTERS...173
GLOBAL WARMING WILL INCREASE THE FREQUENCY AND SEVERITY OF HURRICANES...178
HISTORICAL HURRICANE DATA...182
THE CLAUSIUS-CLAPEYRON RELATIONSHIP 187
GLOBAL WARMING WILL INCREASE FLOODING............................188
GLOBAL WARMING WILL INCREASE THE FREQUENCY OF DROUGHTS AND THEIR SEVERITY...190
CHAPTER 7 CONCLUSIONS 192
CHAPTER 7 SUMMARY 193

CHAPTER 8... **195**
SEA-LEVEL RISE AND GLACIERS..195
THE CURRENT RATE OF SEA-LEVEL RISE......................................198
GLACIERS ARE RETREATING 206
GLOBAL WARMING WILL CAUSE EXCESSIVE SEA-LEVEL RISE 210
CHAPTER 8 CONCLUSIONS 217
CHAPTER 8 SUMMARY 218

CONCLUDING REMARKS **219**

ABOUT THE AUTHOR **221**

WORKS CITED **223**

LIST OF TABLES **271**

LIST OF FIGURES **273**

INDEX.. **283**

REVIEWS

Andy May's down-to-Earth approach to the practical aspects of climate change is grounded in a profound knowledge of the science behind it. He explains in easy to understand terms what the evidence about climate change actually means to us, our food supply, our quality of life, and the environment we all love and care about. In the process he exposes the catastrophic alarmist fantasy, based only on untested, error-prone computer models, used to promote a climate state of fear contradicted by the evidence. Andy May proposes that we use our resources to build a world better adapted to the vagaries of climate and the weather instead of wasting them vainly trying to change a climate that does not need fixing. After enjoying the book and following his arguments, I fully agree with him.

---Javier Vinós, PhD. Scientist and Biological sciences researcher.

"... it looked very good".

---Judith A. Curry PhD. Former chair of the School of Earth and Atmospheric Sciences at the Georgia Institute of Technology and noted American climatologist.

FOREWORD

The mystique of professed "climate scientists" is approaching that of a false religion. But it is far from spiritual warfare. Andy May takes on the climatologist priesthood with a barbarian's club of scientific fact. No need to use any finesse when your opponent is just plain wrong.

Andy May and I go way back as far as the field of petroleum exploration and production is concerned. So, I learned to listen when Andy had something to say. *In Climate Catastrophe! Science or Science Fiction?*, Andy lays out the scientific facts and torpedoes the false chants of the global warming hysteria mongers. As you read Andy's book, pay close attention to the graphs of technical and geopolitical data. Then ask the questions:

1. Why haven't I seen this data presented this way before?

2. Who is behind the science fiction of the adverse effects of CO_2 and fossil fuel consumption?

3. What is their goal?

With Andy's book, you will draw your own conclusions and arrive at your own answers to these questions. Now that's Science.

Judith A. Curry in an early review of Andy's book said, "... it looked very good". She should know. Dr. Curry is an American climatologist and former chair of the School of Earth and Atmospheric Sciences at the Georgia Institute of Technology.

Martin Capages Jr. PhD
Author of *The Moral Case for American Freedom*

PREFACE

Some believe that climate science has become hopelessly politicized, and at times this seems true. Observations are often ignored, and models are often used to make dodgy predictions that do not match what we see outside the window.

This book was written in the vain hope that direct observations of climate today and in the past are still relevant. We examine predictions of future man-made climate "catastrophes" considering what we can see and measure today. We hope to convince the objective reader of the following points:

1. Measurements, that is data, come first, before models and predictions, especially predictions from unvalidated models.

2. If you don't see the problem in the data, it's not a problem.

3. Recent warming is not unusual (Chapter 4).

4. Global warming will not destroy the planet or humans, even in the worst projections (Chapter 1).

5. The oceans, the Sun, and the Earth's orbit are the major controls on climate. Humans probably have some effect, but it must be small (Chapter 1 and Chapter 4).

6. Currently global warming is beneficial for the planet and humanity and is likely to be net beneficial for, at least, several more decades (Chapter 2)

7. The time value of money is critical. Spending a lot of money today to fix a possible problem in 100 years is foolish. From

the standpoint of technology development, 100 years might as well be forever (Chapter 3).

8.. Human prosperity leads to a better environment, a healthier population, more adaptability and lower population growth (Chapter 1 and Chapter 2).

9. Poverty leads to a poorer environment, poorer health and higher population growth (Chapter 1 and Chapter 2).

10. Cheap, widely available and reliable energy leads to prosperity (Chapter 1).

11. Cold is worse than hot. Cold weather leads to more deaths and disease, warm weather leads to fewer deaths and less disease (Chapter 6).

12. Humans are adaptable. Today we live in hot areas, cold areas, dry, and wet areas, high in the mountains and in rainforests. We have already adapted, somewhere, to anything foreseen by climate alarmists (Chapter 6).

13. Our food supply is growing rapidly, with no sign of slowing down. Prices are stable. Population growth, on the other hand, is slowing down (Chapter 2).

14. The rate of extinctions today is very low, and we are not in a "great" or "mass" extinction nor are we even close (Chapter 5).

15. The extreme weather trend is flat or declining (Chapter 7).

16. Extreme weather events have a smaller impact today than in the past (Chapter 7).

17. The Gulf Stream is not shutting down (Chapter 5).

18. Our measurements of the rate of sea-level rise are so inaccurate we cannot even be sure that sea level is rising at all, although it probably is at a very slow rate (Chapter 8).

19. Sea-level rise is not alarming, except locally, and should be dealt with as a local problem (Chapter 8).

20. In general, dealing with climate change locally is far more effective and much cheaper than attempting to control the climate globally with fossil fuel controls or taxes (Chapter 3).

In modern times we are often presented with predictions from computer models. The authors of these models sometimes try and present them as if they were facts. They aren't, and they are often wrong. Many computer models predicted Hillary Clinton would be elected President of the United States. Where is she today?

Computer models have confidently predicted that doubling CO_2 will warm the planet by 1.5°C to 4.5°C. This was predicted in 1979 by the National Academy of Sciences "Charney Report" (Charney, et al. 1979), and it was the Intergovernmental Panel on Climate Change (IPCC) AR5 prediction in 2013 (IPCC 2013). After spending over $100 billion in the U.S. alone on thousands of researchers for 34 years, we are still no closer to measuring the effect of man-made CO_2 on our climate. As Dr. Judith Curry has noted, it seems that the uncertainty about the effect of man-made CO_2 on our climate has *increased* since 1979 (Curry 2017).

If the effect is unknown by a factor of three or more, why are we being asked to spend hundreds of billions of dollars to eliminate or reduce fossil fuel use? Proponents of this action readily admit they cannot detect a climate or weather impact of man-made carbon dioxide, but they insist it is an urgent danger. Since we can't see any impact, other than the planet is getting greener due to the CO_2 fertilization effect (Li, et al. 2017) and (Zhu, et al. 2016), how do they justify this urgency? They do it with models, and they do it with models that have yet to successfully predict anything. Thus, the models are "unvalidated."

What we will do in this book is examine their claims using measurements. Do measurements we can make today support their model results or not? Computer models are science fiction, they are reasonable predictions of what might happen in the future with what we know today. But, they are not science. Science is rooted in observations. If we make a prediction that is later verified with measurements, we have a proper scientific theory. A prediction, no matter how elaborately it was made or documented, that is not verified with data and observations is science fiction.

The most important thing I learned while researching this book is that the effects of climate change are ultimately local and must be dealt with locally. Humans live happily all over the world, high in the mountains, below sea level, in temperatures of 50°C (122°F) and -50°C (-58°F). Why not adapt to future climate changes? We have always adapted in the past, why not now? Measuring the impact of climate change, whether natural or man-made, must be done in each community and suited to their special needs. The idea that we need a global uniform response to climate change, with what we know today, is silly and unreasonable.

CHAPTER 1
Do Humans Harm the Environment?

Most of this book will be a discussion of the purported hazards of climate change and global warming. The authority on climate change and the dangers thereof is usually considered to be the Intergovernmental Panel on Climate Change, also called the IPCC. So, we refer to the <u>IPCC WGII AR5 Technical Summary</u> (Field, et al. 2014) definition of "hazards" on page 39:

"The potential occurrence of a natural or human-induced physical event or trend or physical impact that may cause loss of life, injury, or other health impacts, as well as damage and loss to property, infrastructure, livelihoods, service provision, ecosystems, and environmental resources. In this report, the term hazard usually refers to climate-related physical events or trends or their physical impacts."

We will adopt this definition, but before we discuss the specific hazards of global warming and man-made climate change, we should examine the claims that man is or can be an existential threat to either humans or the planet. Do humans harm the environment just by living in it? If we assume humans are causing most of the current global warming, is the warming dangerous? If we are dangerous to the environment, should we limit our population in some way? If global warming is potentially dangerous, and we assume human CO_2 emissions are the cause, would we be better off to adapt to the human-caused global warming and continue using fossil fuels, or do we need to stop using fossil fuels? We will consider these issues here and in future chapters.

In this chapter, we will deal with the more extreme claims. Some claim humans are dangerous, we breed too much, we use too many resources, we are an existential threat to ourselves and the rest of the world. So, before we get into the economic costs and hazards of climate change, let's discuss the more extreme "existential" potential threats.

Can global warming destroy the Earth?

The existential threat is often explained as the Earth will become like Venus, with an average surface temperature of about 460°C to 477°C (NASA 2013) (Williams 2015) [or 250°C, as Stephen Hawking once incorrectly asserted (Ghosh 2017)] and barren of life. James Hansen once called (Hansen 2009) (May 2016) this the runaway greenhouse effect. His idea was that adding fossil fuel CO_2 to the atmosphere will cause warming that will accelerate in an out of control fashion. Venus has an atmosphere that is 97 percent carbon dioxide and is very hot, so Hansen speculated that the carbon dioxide made Venus "runaway."

The truth is that neither the Earth nor Venus are "runaway" (May 2016) (Billings 2013). Further, the Earth has oceans, and Venus has almost no water. 99.9 percent of the Earth's heat capacity and thermal energy is in our oceans. Less than 0.1 percent of the Earth's thermal energy is stored in the atmosphere. The Earth's surface has five times more stored thermal energy than the surface of Venus (May 2016) (D. Williams 2016). The Earth's oceans alone store more thermal energy than the whole surface of Venus with a temperature of over 460°C. If our oceans continue to exist, there is no way our planet's surface could reach a dangerous temperature according to James Kasting at Pennsylvania State University (Billings 2013). They would have to completely boil away, and the water vapor would have to be ejected to outer space. No greenhouse gas could ever accomplish that.

Thought experiment: If the atmosphere could somehow reach a temperature of 1,000°C, lose none of the thermal energy to outer space, and transfer all of it to the oceans; the temperature of the oceans would increase one degree. This is the easiest way I can think of to explain the temperature buffering effect of the oceans. For another, more complete description of how the tropical oceans limit the surface temperature of the Earth to a maximum of 30°C, see (Newell and Dopplick 1979), (Sud, Walker and Lau 1999) or the discussion in (May 2017).

Atmospheric temperatures, especially the temperature of the atmosphere at the surface (basically the lower two meters of the atmosphere) have very little impact on long-term (meaning decades or longer) climate. Attempts to measure the average surface temperature (the HADCRUT (Morice 2017) database, GISTEMP (Schmidt 2017a), etc.) are useful, after all we live on the surface; but using them to measure the impact or severity of global climate change is like measuring the impact of a bomb blast by counting the ripples in a tea cup in a basement TV room 100 kilometers away. The ripples may be related to the blast, but you are too far away from the main event to be accurate. The oceans cover 70 percent of the Earth's surface and contain nearly all the thermal energy; the focus should be on them.

The oceans are key

The core assumption of the CO_2 hypothesis of man-made global warming is that the atmosphere controls climate. Given the distribution of thermal energy on the surface of the Earth, this makes little sense. It is far more likely that the oceans drive climate. As Javier Vinós wrote in 2018:

"There is a significant possibility however that the climate is actually ocean-driven, directly forced by the Sun, and mediated by H_2O changes of state." (Vinós 2018)

Because the oceans cover 70 percent of the Earth's surface, they receive most of the solar thermal energy striking the planet. When the ocean surface reaches a critical temperature of 28°C to 29°C (in the tropics) the ocean can charge the lower atmosphere with enough energy to push clouds to the upper troposphere. The clouds then reflect more solar radiation, the water vapor condenses, releasing latent thermal energy so it can be radiated to outer space, and the resulting cool, dry air plunges downward to further promote ocean cooling (Sud, Walker and Lau 1999). This process is called "deep convection," and it is an efficient way for the planet to eject excess heat. It also acts as a powerful thermostat, limiting the upper temperature of the Earth's surface.

While the temperature required to cause deep convection in the tropics is 28°C to 29°C, the process can be initiated at a lower temperature in the mid-latitudes. It can start in the North Atlantic at temperatures in the 20°C range (Sud, Walker and Lau 1999).

Ocean temperatures do appear to be rising. But, we only have good global data since 2004 and only to a depth of 2,000 meters. The average ocean depth is 3,688 meters. If we assume the temperature at 3,688 meters is about 0°C, which is not unreasonable since the worldwide average temperature at 2,000 meters is 2.4°C, with a standard deviation of only 0.008°C, then we currently see 0.0031°C of warming for the oceans per year to that depth (see Figure 1A). If this continues (it won't) then it would take 1,000 years for the surface temperature of the Earth to increase 3°C, hardly alarming. More on the JAMSTEC ocean temperature grid in (May 2016a).

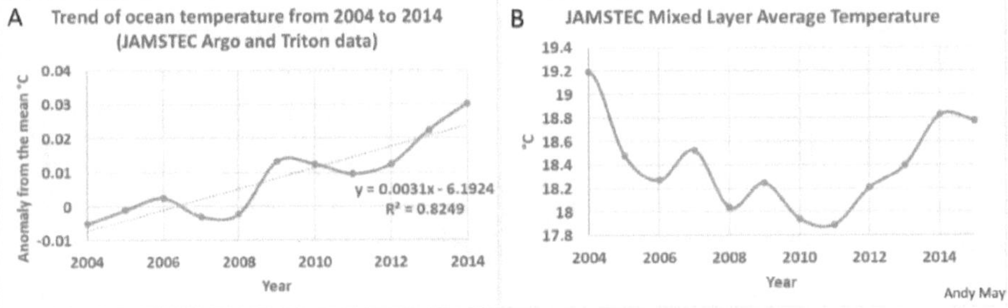

Figure 1. Average ocean temperature and average ocean mixed layer temperature, data sources: JAMSTEC (May 2016a) (JAMSTEC 2016) (JAMSTEC 2016a).

The "mixed layer" of the ocean, according to the JAMSTEC mixed layer data grid, globally, averages about 59.5 meters thick, ±1.3 meters. This is the uppermost layer of the ocean, it is well-mixed and has a vertically uniform temperature, density and salinity. This is the ocean layer that interacts with the atmosphere. The global average temperature of this layer, from the JAMSTEC MILA grid, is plotted in Figure 1B. Over the 11 years plotted, it doesn't appear to have much of a trend, but through 2015 it appears to be declining. However, this is a very short time period, and any trend could change radically with 20 or 30 more years of data.

In future chapters, we will discuss the economic costs of mitigating warming by reducing or eliminating fossil fuel use versus adapting to the warming. In this chapter, we will discuss human impacts on the environment in general. Humans are part of nature, we may have evolved naturally or been created by a supernatural being when the Earth was created, either way we are a part of nature. Deepak Chopra has discussed the "Gaia hypothesis" (Chopra 2005) and wonders if we are a cancer on the Earth. Sir David Attenborough has asserted on BBC that we are a <u>plague</u> on the Earth (Gray 2013).

In this chapter, we examine this idea. We need to get past the idea that man may be an existential threat to man before we discuss the economics of global warming, otherwise some will say: "So, what if adapting to global warming is cheaper, if we are all going to die!" Some of this discussion depends upon point-of-view. Do we take the humanist view that our actions should help mankind? Or, is some sort of metaphysical "Gaia" god-like creature supreme, and mankind must take second place and serve Gaia? Do we really exert that much control over the Earth? We will argue our points as a humanist devoted to the betterment of mankind.

Economic growth, prosperity, and the environment

"In general, we need to confront our myth of the economy undercutting the environment. We have grown to believe that we are faced with an inescapable choice between higher economic welfare and a greener environment. But surprisingly and as will be documented throughout this book, environmental development often stems from economic development – only when we get sufficiently rich can we afford the relative luxury of caring about the environment. On its most general level, this conclusion is evident in Figure [2], where higher income in general is correlated with higher environmental sustainability." Lomborg, Bjørn. *The Skeptical Environmentalist: Measuring the Real State of the World* (pp. 32-33) (B. Lomborg 2001).

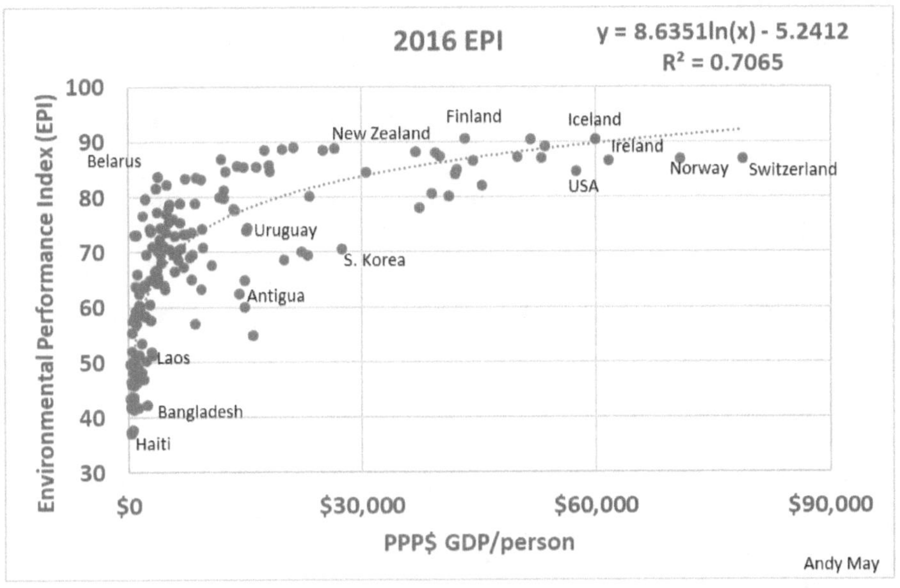

Figure 2. 2016 GDP/person plotted against the NASA environmental performance index, data sources: Environmental Performance Index: (NASA 2017), GDP in PPP$: (World Bank 2017).

Figure 2 is an updated version of a similar figure in *The Skeptical Environmentalist* (B. Lomborg 2001), we have plotted the NASA measure of environmental quality (EPI) for 2016 versus the World Bank 2016 purchase-power-parity dollar (PPP$) GDP for most countries in the world. Some of the countries are labeled. Because different countries have different currencies and the cost of goods varies from country to country, the World Bank converts all local currencies to PPP$ to make valid comparisons between countries (Investopedia 2016). PPP$ is the true exchange rate between countries, adjusted for the cost of a basket of goods from each country. Over time it is also adjusted for inflation.

The plot contains points for 164 countries, data was unavailable for many countries and Luxembourg was excluded because of a very high GDP/person ($102,831, EPI=86.6, about the same as Switzerland). A logarithmic least squares line through the data is a decent fit with an R^2 of 0.7, even though there are many other factors affecting both GDP and EPI for all the countries plotted. So, Lomborg was correct in his 2001 book; the wealthier countries tend to have higher environmental quality. One can also be convinced by visiting developing countries and developed countries around the world. Once GDP/person exceeds about PPP$2,000/person, affluence begins to become a factor in environmental quality until it flattens out in the high 80s at around PPP$50,000/person.

Purchase power parity GDP and human development also correlate well with energy consumption (ExxonMobil 2017) (May 2017b). This suggests that affluence, critical to protecting the environment, is highly dependent upon access to cheap and abundant energy. Fossil fuels, nuclear and hydroelectric are our cheapest and most abundant sources of energy (Weißbach, et al. 2013). They have led to our current prosperity, and they allow us to care for the environment. Without them will the environment improve or get worse? It is important that we always consider the consequences of eliminating fossil fuels. Will it accomplish what we expect? What else will happen?

Disease and health

"In October 1998, Professor [David] Pimentel [Cornell University] published as lead author an article on the "<u>Ecology of increasing disease</u>" (Pimentel, Tort, et al. 1998) in the peer-reviewed journal *BioScience*. The basic premise of the paper is that increasing population will lead to increasing environmental degradation, intensified pollution and consequently more human disease." Lomborg, Bjørn. *The Skeptical Environmentalist: Measuring the Real State of the World* (p. 22) (B. Lomborg 2001).

In a 2007 paper, <u>Pimentel, et al.</u> (Pimentel, Cooperstein, et al. 2007) double down:

> "The World Health Organization (WHO) and other organizations report that the prevalence of human diseases during the past decade is rapidly increasing. Population growth and the pollution of water, air and soil are contributing to the increasing number of human diseases worldwide. Currently an estimated 40 percent of world deaths are due to environmental degradation. The ecology of increasing diseases has complex factors of environmental degradation, population growth and the current malnutrition of about 3.7 billion people in the world."

The "40 percent of world deaths" in the quote above is cited in the 2007 paper as being from Pimentel's 1998 paper referred to in the Lomborg quote. Lomborg has a lot to say about Pimentel's 1998 paper. First, the "40 percent of world deaths" is never explained in the paper, neither the total number of deaths nor the number due to pollution are specified in the article. Deeper in the article the reason changes from "pollution" to "pollution, tobacco and malnutrition." In a later interview he explains that "smoking" includes burning wood in the home. According to Pimentel's source, the World Health Organization (WHO), in the third world burning biomass (mainly wood) in the home kills 4 million people a year (WHO 2017), and smoking kills 7 million a year (WHO 2017b). In 2013, malnutrition was linked to the deaths of three million children per year according to WHO (WHO, UNICEF, et al. 2014). WHO estimated in 2014 that outdoor air pollution kills about 2.6 million people per year (WHO 2014c), which is less than half of the total. This data is summarized in Table 1. The causes listed sum to almost 30 percent, in the ballpark of 40 percent, depending upon the year the WHO statistics were gathered.

World Health Organization Worldwide Deaths		
2017		
	millions of people	%
Total Deaths	56.4	
Indoor air pollution	4	7.1%
Outdoor air pollution	2.6	4.6%
Child Malnutrition	3	5.3%
Smoking	7	12.4%

Table 1. Worldwide Deaths from the World Health Organization (WHO).

In a separate 2016 report, the World Health Organization estimates that 12.6 million deaths (WHO 2016) are due to either air (indoor or outdoor and including secondhand tobacco smoke), water, and soil pollution, chemical exposure, climate change, or excessive ultraviolet radiation. This amounts to 22 percent of total deaths. 8.2 million of the 12.6 are due to indoor or outdoor air pollution. Water and soil pollution-related deaths are declining as access to safe water and proper sanitation improves. So, Pimentel, et al., in both papers, have taken a small number of pollution-related deaths, added smokers, the malnourished, people who cook and heat their houses with wood to that number and have tried to claim they all died due to environmental pollution. Who were the peer-reviewers?

The main claim of the paper, that disease is increasing, is also incorrect.

"The claim about increasing infectious disease is downright wrong, as can be seen in [Figure 3]. Infectious diseases have been decreasing since 1970 and probably much longer, though we only have evidence from some countries … Likewise infectious disease is expected to decrease in the future, at least until 2020. Even in absolute numbers, infectious deaths are expected to drop from 9.3 million to 6.5 million." Lomborg, Bjørn. The Skeptical Environmentalist: Measuring the Real State of the World (p. 26).

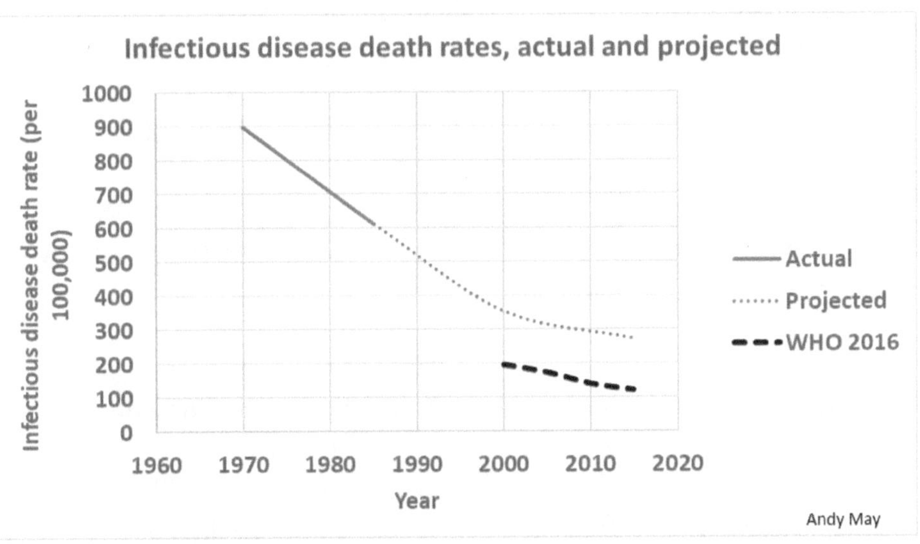

Figure 3. World infectious disease death rates, by year, data sources: blue curves (Bulatao 1993) and the black heavy dash curve, (WHO 2016b).

The data shown in Figure 3 are also <u>supported</u> by (Murray and Lopez 1997). The blue solid curve in Figure 3 is from data compiled by Bulatao and published in 1993. The black, heavy, dashed curve is of data compiled from 186 countries by the World Health Organization (WHO) and published in December 2016. Bulatao's projection (the light, dashed line) got the trend right but was a little pessimistic. The rate of infectious disease dropped faster than he projected. So, again, Lomborg was correct.

"When looking at trends, Pimentel happily uses very short-term descriptions. He looks at the biggest infectious disease killer, tuberculosis, claiming it has gone from killing 2.5 million in 1990 to 3 million in 1995, and citing an expected 3.5 million dead in 2000. However, in 1999, the actual death toll from tuberculosis was 1.669 million, and the WHO source that Pimentel most often uses estimates an almost stable 2 million dead over the 1990s." Lomborg, Bjørn. *The Skeptical Environmentalist: Measuring the Real State of the World* (pp. 22-23).

According to the WHO 2016 statistics, there were 1,667,000 deaths due to tuberculosis in 2000 and 1,373,000 in 2015. This is a drop of 18 percent in 15 years.

"Equally, pointing out the danger of chemicals and pesticides, Pimentel tries to make a connection by pointing out that "in the United States, cancer-related deaths from all causes increased from 331,000 in 1970 to approximately 521,000 in 1992." However, this again ignores an increasing population (24 percent) and an aging population (making cancers more likely). The age-adjusted cancer death rate in the U.S. was actually lower in 1996 than in 1970, despite increasing cancer deaths from past smoking, and adjusted for smoking the rate has been declining steadily since 1970 by about 17 percent." Lomborg,

Bjørn. *The Skeptical Environmentalist: Measuring the Real State of the World* (p. 23).

The cancer death rate did increase from 1972 to 1990 according to the (CDC 2017), but after 1990 it fell dramatically as can be seen in Figure 4. The values plotted in Figure 4 are age-adjusted by the CDC.

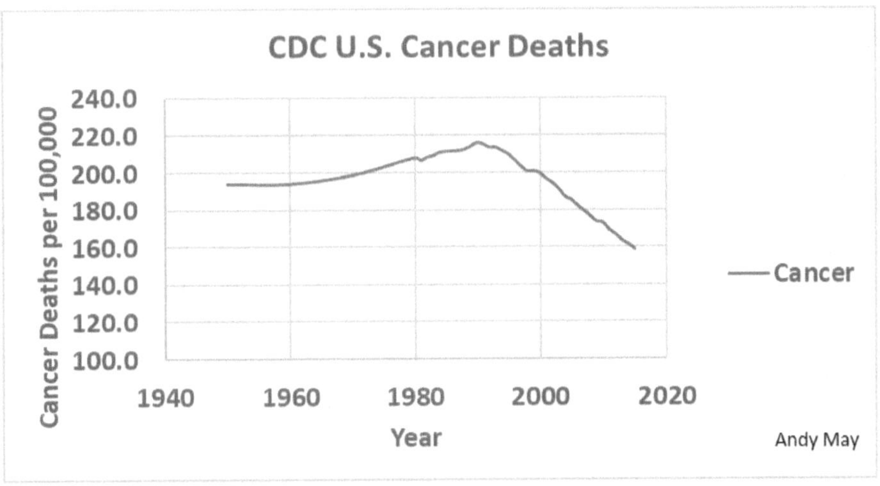

Figure 4. Age adjusted U.S. death rate due to cancer, data from (CDC 2017).

In this case, Pimentel is correct that cancer deaths increased from 1970 to 1992, even when adjusted for age and population. However, Lomborg is also correct that the adjusted cancer death rate in 1996 was lower than in 1970. It is much lower today.

"In 2002 the World Health Organization concluded that it could not identify the influence of greenhouse gas emissions on health and disease based on existing data: "Climate exhibits natural variability, and its effects on health are mediated by many other determinants. There are currently insufficient high-quality, long-term data on climate-sensitive diseases to provide a direct measurement of the health impact of anthropogenic climate change, particularly in the most vulnerable populations." Pielke Jr., Roger. *The Climate Fix: What Scientists and Politicians Won't Tell You About Global Warming* (pp. 176-177) (Pielke Jr. 2010).

"The speculative guesses of WHO formed the basis of estimates released in a 2009 report issued by the Global Humanitarian Forum, a non-governmental organization run by former UN secretary-general Kofi Annan. The GHF concluded that greenhouse gas-driven climate change was presently responsible for 154,000 deaths per year due to malnutrition, 94,000 deaths per year due to diarrhea, and 54,000 deaths per year due to malaria, which when added to deaths from weather-related disasters (which have declined dramatically over the past century) gives a total of 315,000 people who allegedly die each year due to human-caused climate change. A close look at the health-related numbers shows that they are exactly two times the values presented in the 2002 WHO report, which said that the estimates do not "accord with the canons of empirical science." In other words, the numbers appear to be just a guess on top of the earlier speculation. "Analyses" such as these are what give some areas of climate science a bad name and suggest an unhealthy politicization of research to support favored causes." Pielke Jr., Roger. *The Climate Fix: What Scientists and Politicians Won't Tell You About Global Warming* (p. 177) (Pielke Jr. 2010).

Regarding the health effects of global warming, the IPCC WGII AR5 Technical Summary has this to say on page 71 (Field, et al. 2014):

"Until mid-century, projected climate change will impact human health mainly by exacerbating health problems that already exist (very high confidence). Throughout the 21st century, climate change is expected to lead to increases in ill-health in many regions and especially in developing countries with low income, as compared to a baseline without climate change (high confidence). Examples include greater likelihood of injury, disease and death due to more intense heat waves and fires (very high confidence); increased likelihood of under-nutrition resulting from diminished food production in poor regions (high confidence); risks from lost work capacity and reduced labor productivity in vulnerable populations; and increased risks from food- and water-borne diseases (very high confidence) and vector-borne diseases (medium confidence). Impacts on health will be reduced, but not eliminated, in populations that benefit from rapid social and economic development, particularly among the poorest and least healthy groups (high confidence)."

Clever wording makes this quote mostly true, but very misleading. As we will see, global warming may increase heat-related deaths, but cold-related deaths will be reduced by a much larger amount. This is because there are more cold-related deaths in the world than heat-related deaths. This is discussed in more detail in Chapter 6. We have already shown above that infectious diseases and cancers are decreasing. In addition, malnutrition, or "under-nutrition," is also decreasing at a rapid rate. Thus, the models that were used to make these projections have yet to be validated and are contradicted by the current data.

Life expectancy

Since 100 percent of us die of something, perhaps a better measure of human health is average life expectancy, which is currently 71.5 years globally. According to the_(World Bank 2017), life expectancy globally is <u>increasing</u> (see Figure 5). According to *The Lancet* (Landrigan 2017)) environmental pollution is responsible for 16 percent of global deaths or nine million premature deaths globally. Ninety-two percent of these premature deaths occur in low- and middle-income countries. This number is slightly lower than the WHO number cited above, but much lower than Pimentel, 2007.

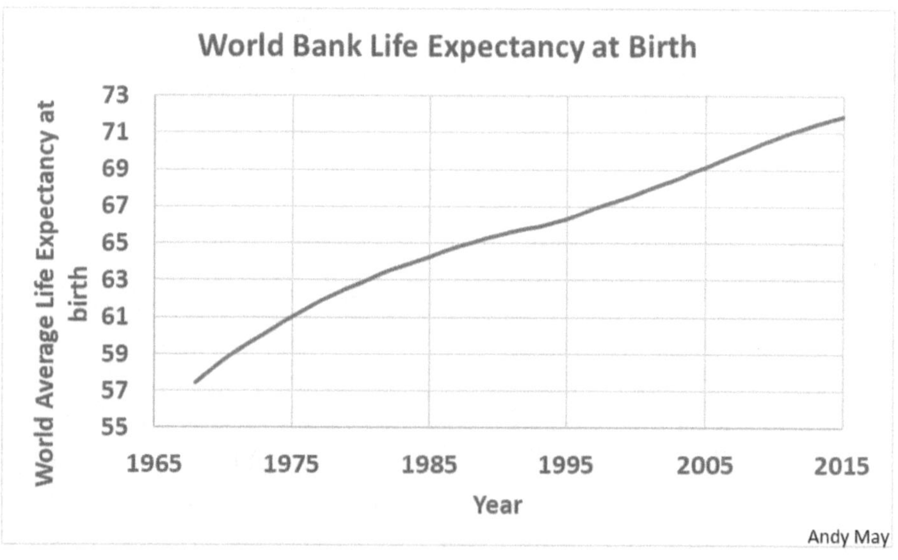

Figure 5. World life expectancy at birth by year, data source: (World Bank 2017).

There is also a correlation between per capita energy use and life expectancy, as you can see in Figure 6. Figure 6 plots worldwide energy use per person in kilograms of oil equivalent versus life expectancy at birth, as estimated by the World Bank. As energy use goes up so does life expectancy.

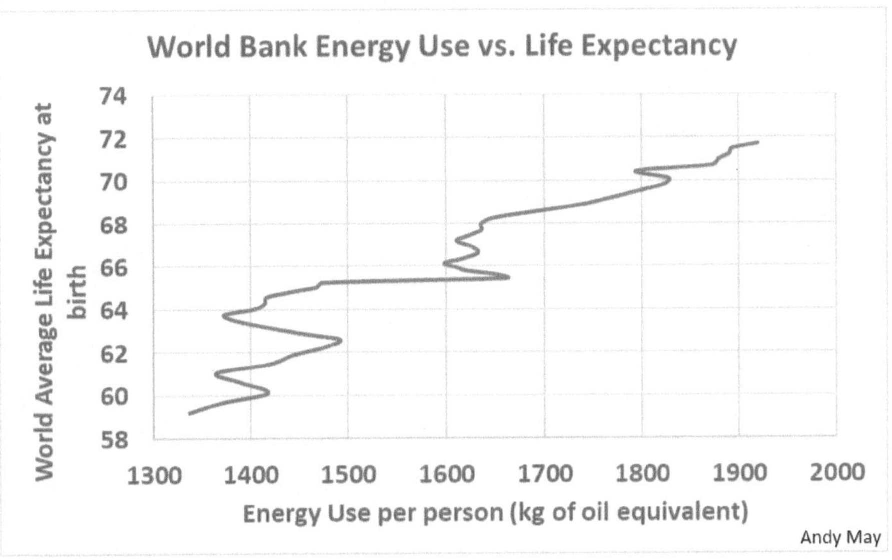

Figure 6. World Bank Energy used per person versus World Bank life expectancy, data source: (World Bank 2017).

Interestingly, as people become more affluent, the energy used to generate a dollar of GDP goes down, that is affluent people are more efficient in their use of energy as can be seen in Figure 7. In Figure 7 the solid line is for the whole world, and the dashed is only for the upper middle class or wealthier individuals. At almost every point, the wealthier people get more GDP per kilogram of oil. Prosperity not

17

only helps people live longer and better lives, it also makes them more efficient with the energy they have access to.

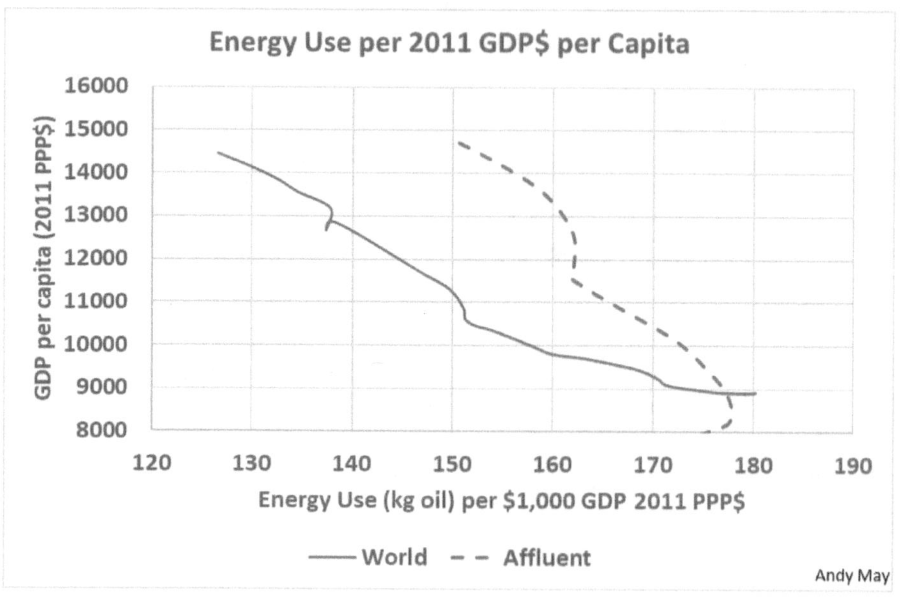

Figure 7. Energy used per person versus GDP per person for 2011. The whole world is compared to affluent countries, data source: (World Bank 2017).

Not only are people living longer, they are also living better lives:

"That is all very well, say pessimists, but what about quality of life in old age? Sure, people live longer, but only by having years of suffering and disability added to their lives. Not so. In one American study, disability rates in people over 65 fell from 26.2 percent to 19.7 percent between 1982 and 1999 — at twice the pace of the decrease in the mortality rate. Chronic illness before death is, if anything, shortening slightly, not lengthening, despite better diagnosis and more treatments — 'the compression of morbidity' is the technical term. People are not only spending a longer time living, but a shorter time dying." Ridley, Matt. *The Rational Optimist: How Prosperity Evolves* (P.S.) (p. 18) (Ridley 2010).

"We live longer, but have we only been given more time in which to be ill? The answer has to be: absolutely not. We have generally become much healthier during the past centuries." Lomborg, Bjørn. *The Skeptical Environmentalist: Measuring the Real State of the World* (pp. 54-55).

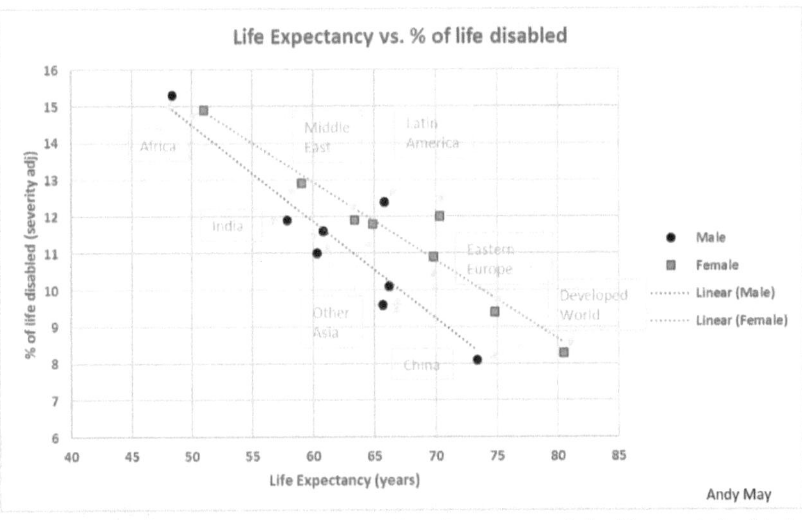

Figure 8. Life expectancy is compared to the percent of the life spent disabled for different areas of the world, data source: (Murray and Lopez 1997-2).

In Figure 8, we see that the developed world has the longest life span at birth and the smallest percentage of their lives spent as disabled. The data are from <u>Murray and Lopez 1997-2</u>, Table 4, page 1350. The disabilities shown on the y-axis are severity-adjusted. As people become wealthier and as they use more energy, they not only live longer, but they live healthier lives. The idea that people in the pre-industrial era lived happy, healthy lives in harmony with nature is nonsense, as explained by Princeton historian Lawrence Stone in <u>The Family, Sex and Marriage in England 1500-1800</u> (Stone 2013):

"[Lomborg quoting Stone] The almost total ignorance of both personal and public hygiene meant that contaminated food and water was a constant hazard ... The result of these primitive sanitary conditions was constant outbursts of bacterial stomach infections, the most fearful of all being dysentery, which swept away many victims of both sexes and of all ages within a few hours or days. Stomach disorders of one kind or another were chronic, due to poorly balanced diet among the rich, and the consumption of rotten and insufficient food among the poor. The prevalence of intestinal worms ... were a slow, disgusting and debilitating disease that caused a vast amount of human misery and ill health ... In the many poorly drained marshy areas, recurrent malarial fevers were common and debilitating diseases ... [and] perhaps even more heartbreaking was the slow, inexorable, destructive power of tuberculosis ... For women, childbirth was a very dangerous experience ... [and finally] there was the constant threat of accidental death from neglect or carelessness or association with animals like horses – which seem to have been at least as dangerous as automobiles." Lomborg, Bjørn. *The Skeptical Environmentalist: Measuring the Real State of the World* (p. 55).

Technology exists to control both air and water pollution, and it is rapidly spreading to poor- and middle-income countries. In the U.S. where the technology has been used for decades, air pollution has <u>decreased 70 percent</u> (EPA 2017) since 1970. According to UNICEF,

2.6 billion people have gained access to an improved drinking water source since 1990 (UNICEF and WHO 2017).

"Note that as recently as 1990, under half the world had "improved sanitation facilities." The increase to two-thirds in only a few decades is a wonderful accomplishment," Epstein, Alex. *The Moral Case for Fossil Fuels* (p. 148). Penguin Publishing Group. Kindle Edition. (Epstein 2014)

And:

"The rich have got richer, but the poor have done even better. The poor in the developing world grew their consumption twice as fast as the world as a whole between 1980 and 2000. ... Roughly eight out of ten American households had running water, central heating, electric light, washing machines and refrigerators by 1955. Almost none had these luxuries in 1900. ... Today, of Americans officially designated as 'poor', 99 percent have electricity, running water, flush toilets, and a refrigerator; 95 percent have a television, 88 percent a telephone, 71 percent a car and 70 percent air conditioning." Ridley, Matt. *The Rational Optimist: How Prosperity Evolves* (p. 15-17). (Ridley 2010)

In some ways, the poor in the U.S. are better off than King Louis XIV of France in 1700, the richest man of his day. The poor today own goods and eat food he could only dream of.

Is the environment deteriorating?

Perhaps in developing countries, but:

"In Europe and America rivers, lakes, seas and the air are getting cleaner all the time. The Thames has less sewage and more fish. Lake Erie's water snakes, on the brink of extinction in the 1960s, are now abundant. Bald eagles have boomed. Pasadena has few smogs. Swedish birds' eggs have 75 percent fewer pollutants in them than in the 1960s. American carbon monoxide emissions from transport are down 75 percent in twenty-five years. Today, a car emits less pollution traveling at full speed than a parked car did in 1970 from leaks." Ridley, Matt. *The Rational Optimist: How Prosperity Evolves* (p. 17).

Prosperity leads to a cleaner environment. Once people have food, energy and shelter, they turn to cleaning up their environment.

Prosperity

People now live much longer and better lives than 100 years ago. Conveniences, such as artificial light, are much cheaper today:

"[Consider] how much artificial light you can earn with an hour of work at the average wage. The amount has increased from twenty-four lumen-hours in 1750 BC (sesame oil lamp) to 186 in 1800 (tallow candle) to 4,400 in 1880 (kerosene lamp) to 531,000 in 1950 (incandescent light bulb) to 8.4 million lumen-hours today (compact fluorescent bulb). Put it another way, an hour of work today earns you 300 days' worth of reading light; an hour of work in 1800 earned you ten minutes of reading light." Ridley, Matt. *The Rational Optimist: How Prosperity Evolves* (P.S.) (pp. 20-21).

In the mid-1800s a stagecoach ride from Paris to Bordeaux cost a month's wages. Today a train ticket is 10 Euros. In 1840 transporting a family of four along the Oregon Trail from St. Louis, Missouri, to Oregon City, Oregon, cost $23,373 in 2016 U.S. dollars, and it took 4.5 months to make the trip. Yet, today the car trip is only $1,052, and the 4-hour plane trip is only $1016 (May 2015).

We benefit from what Matt Ridley calls the multiplication of labor. Each of us works at one job, producing one thing; but we purchase goods from all over the world that involved the labor of thousands of people. We can do this because of fossil fuels, modern transportation and win-win capitalism where both the buyer and the seller profit from each transaction.

Consider this story:

"Kelly Cobb of Drexel University set out to make a man's suit entirely from materials produced within 100 miles of her home. It took twenty artisans a total of 500 man-hours to achieve it and even then, they had to get 8 percent of the materials from outside the 100-mile radius. If they worked for another year, they could get it all from within the limit, argued Cobb." Ridley, Matt. *The Rational Optimist: How Prosperity Evolves* (P.S.) (p. 35).

500 man-hours of skilled labor to make a suit! In the U.S. the labor costs alone would be over $5,000. Compare a modern woman in Paris to King Louis XIV, who had 498 servants preparing each meal:

"[A] woman of 35, living in, for the sake of argument, Paris and earning the median wage, with a working husband and two children. You are far from poor, but in relative terms, you are immeasurably poorer than Louis was. Where he was the richest of the rich in the world's richest city, you have no servants, no palace, no carriage, no kingdom. As you toil home from work on the crowded Metro, stopping at the shop on the way to buy a ready meal for four, you might be thinking that Louis XIV's dining arrangements were way beyond your reach. And yet consider this. The cornucopia that greets you as you enter the supermarket dwarfs anything that Louis XIV ever experienced (and it is probably less likely to contain salmonella). You can buy a fresh, frozen, tinned, smoked or pre-prepared meal made with beef, chicken, pork, lamb, fish, prawns, scallops, eggs, potatoes, beans, carrots, cabbage, aubergine, kumquats, celeriac, okra, seven kinds of lettuce, cooked in olive, walnut, sunflower or peanut oil and flavoured with cilantro, turmeric, basil or rosemary ... You may have no chefs, but you can decide on a whim to choose between scores of nearby bistros, or Italian, Chinese, Japanese or Indian restaurants, in each of which a team of skilled chefs is waiting to serve your family at less than an hour's notice. Think of this: never before this generation has the average person been able to afford to have somebody else prepare his meals." Ridley, Matt. *The Rational Optimist: How Prosperity Evolves* (pp. 36-37).

Louis XIV, the Sun King, had 498 servants prepare his meals; yet, he was not as rich as the modern Parisian woman. This woman has the products of thousands of workers at her fingertips in any grocery store or restaurant.

Chapter 1 Conclusions

"All in all, it must be said that mankind's health situation has improved dramatically over the past couple of hundred years. We live to more than twice the age we did just a hundred years ago, and the improvement applies to both the industrialized and the developing world. Infant mortality has fallen in both developed and developing countries by far more than 50 percent." Lomborg, Bjørn. *The Skeptical Environmentalist: Measuring the Real State of the World* (pp. 58-59).

"… there has been a 36-fold increase in per capita American production since 1789, and a similar 20-fold British increase since 1756. In 2000 the U.S. economy produced goods and services for an average American at the value of $36,200; at the end of the eighteenth century, an American would have made just 996 present-day dollars. The average Briton had £15,700 in 2000 compared to just 792 present day pounds in 1756." Lomborg, Bjørn. *The Skeptical Environmentalist: Measuring the Real State of the World* (p. 70).

Subsequent chapters will discuss the costs and benefits of global warming. But, before we get into the economics of global warming, I felt I had to deal with the more extreme "dangerous climate change" claims.

In this chapter, I have tried to discredit three common assertions made by environmentalists. The first assertion is that man is an existential threat to man and the Earth, or "Gaia." The clear answer is "no" to both. Does our growing population cause more infectious disease? Does our economic growth harm the environment? Neither of these statements are true. Rather, prosperity allows us to take better care of our environment and health. Does our growing population and lengthening life expectancy just make us sicker? Clearly not, we live both longer and better lives today. Is the environment getting worse with time? No, in fact it is getting better the more prosperous we are.

Look at the graphs in Figure 9. We discuss them left to right and top to bottom. We plotted a few indicators of trends in human well-being using World Bank data (World Bank 2017). World food production is up over 300 percent since 1968. World poverty, using the World Bank definition of $1.90/day, in constant purchase power parity dollars, is down to 10 percent of the population from over 40 percent in 1980, even though the population has grown. World gross national income per person, in inflation-corrected dollars, is up over 50 percent, and the percentage of children under five years old who die has been cut by more than half since 1990. Even though the population has increased, the number of children, under five, who die has decreased from over 12 million in 1990 to 7 million in 2015.

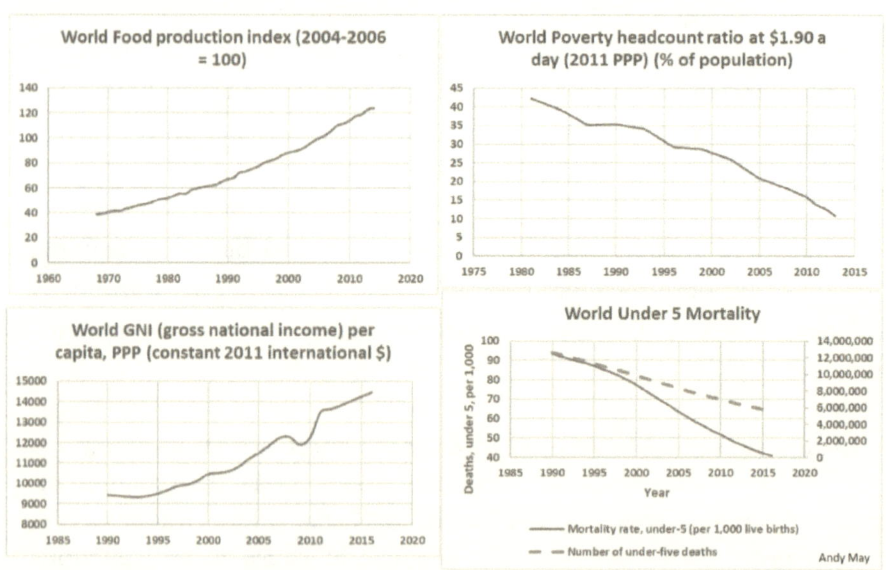

Figure 9. Measures of human prosperity and well-being, data source (World Bank 2017).

It's clear that environment and human welfare are improving as we become more prosperous. Prosperity is the result of new technology, less national control of economies (that is more capitalism), more energy consumption, and more trade.

The second assertion is that global warming has the potential to destroy the Earth by turning it into a Venus-type planet. This is clearly physically impossible if the Earth has oceans. And, no matter how much greenhouse gases increase, the oceans will still be here. The average temperature of the water in the oceans is between 4°C and 5°C and the oceans have 99.9 percent of the heat capacity on the surface of the Earth, so surface warming is severely limited.

The third assertion is that humans are breeding ourselves into extinction or starvation. Are humans outstripping the Earth's resources? No, we are not. Natural resource production and discovery, farm land and food supply are all growing faster than we are using them. And this will be the case for the foreseeable future.

In the strictest sense, I don't have to disprove any of these ideas. They are all very speculative, and there is no data to support any of them, just unvalidated computer models. But, we hear very notable academics, including <u>Stephen Hawking</u> (Ghosh 2017), and politicians repeating this stuff all the time. This chapter is an attempt to inject some reality into the fog of wild speculation.

We are living in a time of nearly boundless prosperity. The rate of poverty has plunged to unimaginable lows. This is a time when the definition of poverty in the United States is set so high, a poor person in the U.S. would be the envy of any middle-class person prior to World War II. Inequality in the world is at its lowest level ever and decreasing at a rapid rate. People who were born in abject poverty can now become doctors and lawyers. Why we still have doomsayers predicting the end of the world is beyond my understanding.

Chapter 1 Summary

"[Humans] don't take a safe climate and make it dangerous; we take a dangerous climate and make it safe." Epstein, Alex. *The Moral Case for Fossil Fuels* (p. 126).

Using fossil fuels, modern transportation, and communication technology, we have built a worldwide system of trade that allows each of us to benefit from the labor of thousands of people around the world with each purchase we make. This has dramatically reduced poverty and worldwide economic inequality. It is also true that more affluent countries have less environmental pollution, with cleaner air, soil and water than less developed countries. This is true, even though richer countries burn more fossil fuels.

Global warming and fossil fuel use will not destroy human civilization or the planet. The oceans will keep any future warming safe and limit the maximum surface temperature. As we will see, all the potential hazards of climate change can be dealt with locally using modern technology. The warming we are observing is mostly global, although as we will see in Chapters 2 and 4, much of the warming is limited to the Northern Hemisphere. But, global or not, the impacts are local and can be ameliorated locally.

Disease is decreasing in the world as people become more prosperous and have more access to safe and clean food, water and energy. The use of "chemicals" and "pesticides" are not causing cancer rates to increase, in fact cancer rates are declining. Life expectancy is increasing worldwide, and the proportion of time an average person spends disabled is decreasing as well.

In this chapter we have tried to establish that human developed technology, which is largely based upon fossil fuels, is not leading to the end of the world or destroying our environment. In fact, abundant cheap energy and our technology have made us prosperous enough to take better care of our environment. In the next chapter, we will explore human population growth and our food supply. Will we outstrip our food supply and starve?

CHAPTER 2

Population Growth and the Food Supply

The IPCC WGII AR5 report only deals with population growth tangentially, but it has a lot of information on the world's food supply. The WGII AR5 technical summary (Field, et al. 2014) states on page 70:

> "For the major crops (wheat, rice and maize) in tropical and temperate regions, climate change without adaptation is projected to negatively impact aggregate production for local temperature increases of 2°C or more above late-20th-century levels, although individual locations may benefit (medium confidence)."

As we will show below, there is no indication of a change in the rate of increase in crop yields in the United Nations Food and Agricultural Organization (FAO) data (FAO Statistics Division 2017). They include the phrase "without adaptation" in their prediction. Since the developing world is adopting the farming technology of the developed world, especially irrigation systems, hydroponics, fertilizer and farm machinery at an accelerating rate, this is good protection in case their "projections" fail. We will also discuss the trends in human population growth.

Population growth

Ever since Thomas Malthus published his <u>theory of population growth</u> (Malthus 1798) there have been large groups who believe that man will breed himself into starvation by consuming all the Earth's resources. We've had over 200 years to prove him correct, but instead today with a population of 7.6 billion, versus the 1798 population of around 800 million, we have more food per person today. In 1960 the world produced 2,200 Calories (a food Calorie is capitalized and is equal to one kilocalorie of energy) per person, similar to what the average person had in the U.K. or France in 1800 (Roser and Ritchie 2018b). Today the world produces 2,900 Calories per person. While no worldwide statistics exist for 1798, about 30 percent of the populations of the U.K. and France were undernourished at the time. Today FAO worldwide statistics show fewer than 11 percent are undernourished (Roser and Ritchie 2018c).

As recently as 1968, Paul Ehrlich published *The Population Bomb* (Ehrlich 1968), which predicted, among other things, that hundreds of millions would starve to death in the 1970s and that England would not even exist in the year 2000. He also warned that the world as we know it will cease to exist before 1985. Well, just like Malthus, Ehrlich was comically wrong.

The world today is improving at a very rapid pace. The only way to see it otherwise is to see humans, as Sir David Attenborough did in 2013, as a "<u>plague on the Earth</u>" (Gray 2013). As for the idea that the human population is growing too fast, consider this:

"As you can see in Figure [10], the massive growth in the world's population began around 1950 and will probably end around 2050. The increase in the population is mainly due to a dramatic fall in the death rate as a result of improved access to food, medicine, clean water and sanitation. The increase is not, on the other hand, due to people in developing countries having more and more children. In the early 1950s women in developing countries gave birth to an average of more than six children – compared to an average of around three today. As [UN Consultant Peter Adamson] put it, rather bluntly: "It's not that people suddenly started breeding like rabbits; it's just that they stopped dying like flies." Lomborg, Bjørn. *The Skeptical Environmentalist: Measuring the Real State of the World* (pp. 45-46). (B. Lomborg 2001)

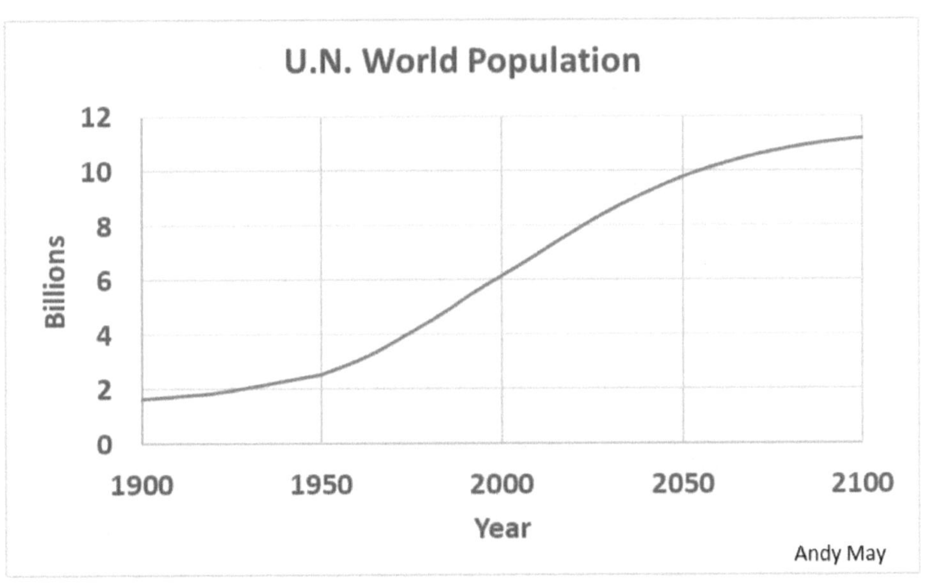

Figure 10. U.N. world population growth, data source: (United Nations Population Division 2017).

"In a traditional agricultural society, income is low and mortality high. However, children working and providing for their parents in old age generally supply greater benefits than their cost, and therefore the birth rate is high. With improved living conditions, medicine, sanitation and general economic prosperity, the death rate falls. The transition toward a more urban and developed economy makes children more likely to survive while they start to cost more than they contribute, needing more education, working less and transferring the care of their parents to nursing homes. Consequently, the birth rate drops." Lomborg, Bjørn. *The Skeptical Environmentalist: Measuring the Real State of the World* (p. 46).

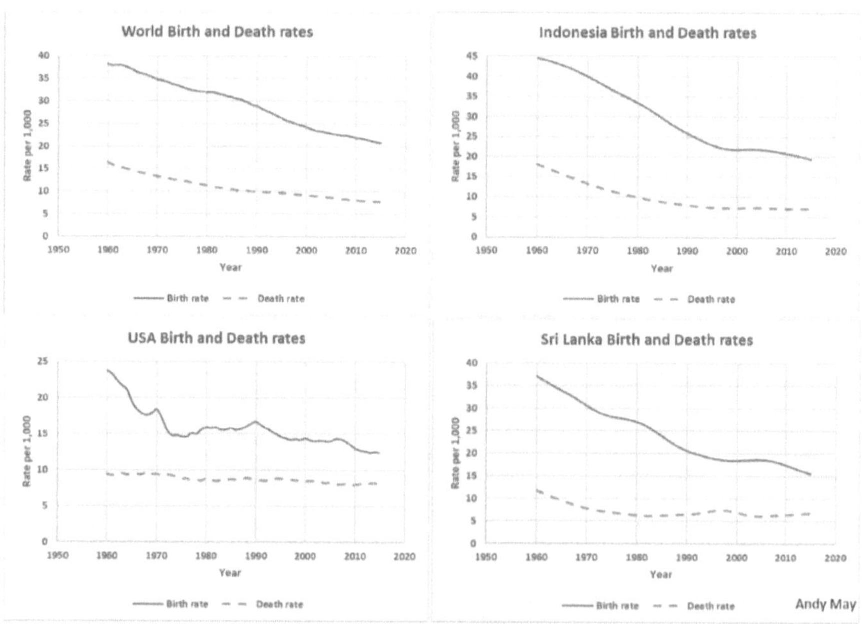

Figure 11. Birth and death rates for four countries, data source: (World Bank 2017).

In Figure 11 we see the death rate falls first and faster than the birth rate. Longer term this leads to longer life expectancy. In Figure 12 we see the U.S. Census Bureau estimates of world population growth in absolute terms and in percent. In absolute terms the growth rate peaked in 1989 at 88 million people a year, it fell to 79 million by 2016 and the rate of population growth is currently 1.1 percent, down from over 2 percent in the early 1960s. It is expected to continue to fall and is projected to be less than 0.1 percent in 2100.

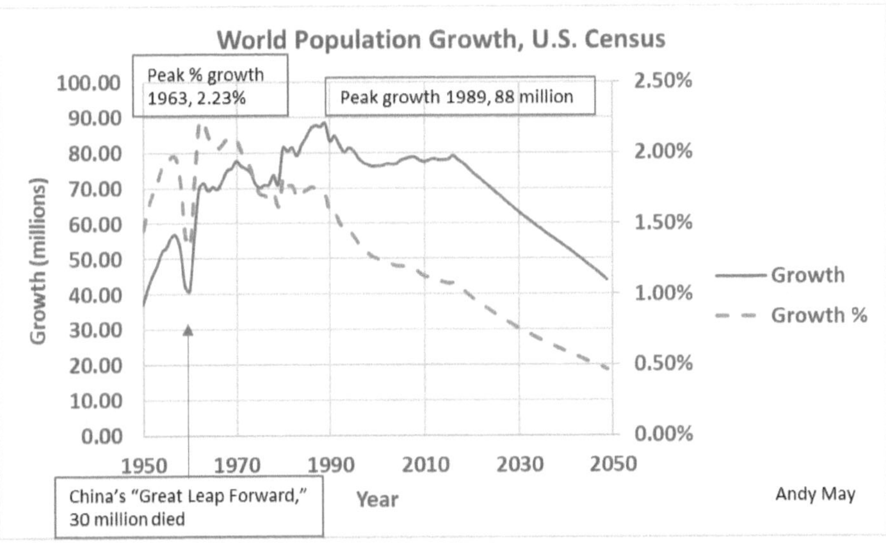

Figure 12. World population growth as a percent of population and in people added, data source: (U.S. Census Bureau 2017).

It's silly to blame population density for a poor quality of life, since the population density of Southeast Asia is the same as in the U.K. The problem with the quality of life in Southeast Asia is not population density, it's poverty.

As the modern industrial era began in the early 1900s, life expectancy increased quickly. Similar improvements occurred in the developing world, although they began later. By 1950 the developing world had a life expectancy of 41 years, and it improved to 65 by 1998. Life expectancy has improved overall, but always correlates with income (see Figure 13).

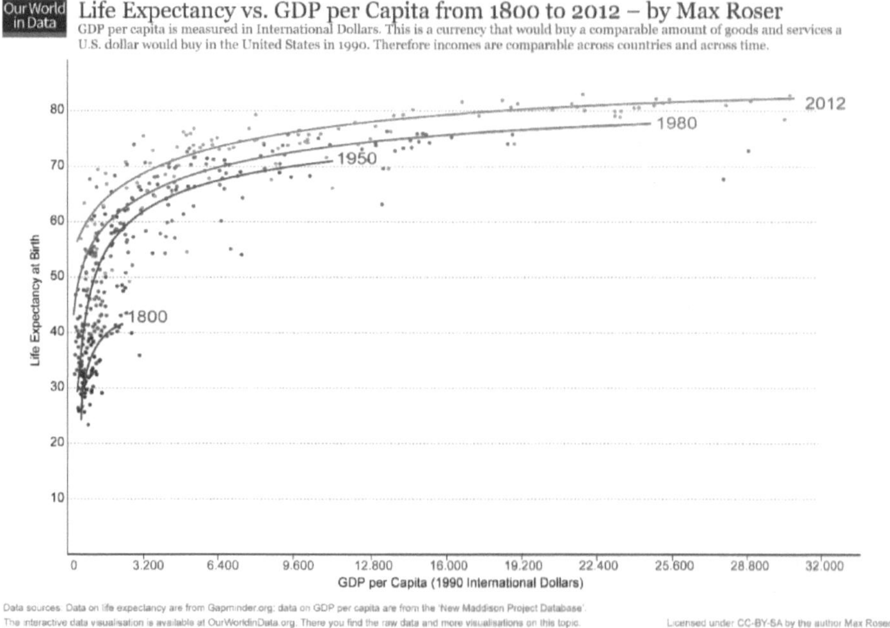

Figure 13. Life expectancy versus GDP per person, data source (Roser, Life Expectancy 2017) used under the CC BY-SA license.

"Professor David Pimentel of Cornell University is a frequently cited and well-known environmentalist, responsible – among many other arguments – for a global erosion estimate far larger than any other ... and for arguing that the ideal population of a sustainable U.S. would be 40–100 million (i.e. a reduction of 63–85 percent of the present population)." Lomborg, Bjørn. *The Skeptical Environmentalist: Measuring the Real State of the World* (p. 22).

According to the <u>FAO</u> (FAO Statistics Division 2017), in 1990 there were one billion malnourished people in the world (18.6 percent) and in 2014 there were 794 million (10.9 percent). A reduction in the number of malnourished, even though the total population had increased to 7.2 billion in 2014 from 5.3 billion in 1990 (36 percent). According to <u>UNICEF</u> (UNICEF 2017b), the number of stunted children has decreased from 254 million in 1990 to 155 million in 2016. The number of overweight children has increased from 32 million to 41 million. There is no evidence that the population must be decreased.

The last major attempt to decrease a population was the one-child policy in China. It has since been rescinded and China has dramatically reduced hunger and malnutrition by using modern technology, trade, and fossil fuels. The life expectancy at birth increased from <u>40 years to 70 years</u> (ChinaPower 2017) between 1950 and 2000. This is due to a reduction in disease and in infant mortality, as well as a decrease in malnutrition. China has <u>decreased malnutrition</u> (World Food Programme 2017) from 24 percent in 1990 to 9 percent in 2016. Child malnutrition has dropped by over 70 percent.

The effect of CO$_2$ and global warming on our food supply

"Stories of how global warming will 'greatly increase the number of hungry people' and of how we are facing 'catastrophe' with 'whole regions becoming unsuitable for producing food' abound." Lomborg, Bjorn. *Cool It* (Kindle Locations 1549-1550) (B. Lomborg 2007)

Consider this story from 1968 by Paul Ehrlich (*The Population Bomb*) (Ehrlich 1968), as quoted from Lomborg's book *The Skeptical Environmentalist*:

"'The battle to feed humanity is over. In the course of the 1970s the world will experience starvation of tragic proportions – hundreds of millions of people will starve to death. … ['professional optimists'] say, for instance, that India in the next eight years can increase its agricultural output to feed some 120 million more people than they cannot after all feed today. To put such fantasy into perspective one need consider only …', and Ehrlich presented a whole list of reasons why this could not be achieved. And sure enough, it turned out that the figure of 120 million did not hold water. Eight years later India produced enough food for 144 million more people. And since the population had grown by 'only' 104 million, this meant there was more food to go around." Lomborg, Bjørn. *The Skeptical Environmentalist: Measuring the Real State of the World* (p. 60).

(Li, et al. 2017) in *Ecological Indicators* in 2016 have reported that the decadal mean net primary plant productivity has <u>increased</u> from 56.12 Petagrams of Carbon (Pg C) to 66.26 Pg C from the 1960s to the 2000s, an increase in net plant productivity worldwide of over 18 percent since 1961. Li, et al. conclude that the increase in atmospheric CO_2 accounts for 45 percent of this increase in productivity. They compute that a doubling of CO_2 to 800 ppm will increase net plant productivity by 60 percent. The Earth has warmed since 1960, and this has allowed plants to enter new areas that were previously too cold, increasing the amount of arable land and increasing the growing season in temperate and Arctic areas. This has contributed to an increase in the Leaf Area Index (LAI). The increase in leaf area contributed 22 percent of the increase in net plant productivity. The other factors considered in the study were climate change and the fraction of photosynthetically active solar radiation available. These two factors contributed 18 percent and 15 percent, respectively, to the increased plant growth. All factors considered were positive for plant growth.

Zaichun Zhu and colleagues, in an article in *Nature Climate Change*, entitled "Greening of the Earth and its drivers," in 2016, conclude that CO_2 fertilization accounts for 70 percent of the observed recent greening of the Earth. Climate change (warming) and landcover change, together, account for an additional 12 percent. Warming stimulates growth in cooler areas and lengthens the growing season. The landcover change is mostly in the northern latitudes, especially in Siberia where previously barren areas are now green. Another area where barren areas have turned green is the Tibetan plateau (Zhu, et al. 2016).

Dr. Craig Idso and others have estimated that the monetary benefit to the world economy of rising CO_2, on global plant production could rise to $10 trillion by 2050 (C. Idso 2017). Idso has calculated that the increase in CO_2 from 1961 to 2011 has already increased crop production value by three trillion dollars (C. Idso 2013). There is overwhelming evidence (see the bibliography of Idso's book and Li, et al., 2017) that increasing atmospheric CO_2 is greatly benefiting farmers worldwide. In addition, western farming technology is spreading around the world very quickly. This also increases farm yields.

"… food availability has increased dramatically over the past four decades. The average person in the developing world has experienced a 40 percent increase in available Calories. Likewise, the proportion of malnourished has dropped from 50 percent to less than 17 percent. The U.N. expects these positive trends to continue at least till 2050 with another 20 percentage points' Calorie increase and malnourished dropping below 3 percent." Lomborg, Bjorn. *Cool It* (Kindle Locations 1552-1556).

Malnourished children, war and climate change

In 2016 the world had 815 million (FAO 2017b) malnourished people, according to an estimate by the FAO. This is an increase from the FAO estimate of 777 million in 2015 after a long steady decline from over 900 million in 2004. The WHO data repository (WHO 2017c) shows the global number of malnourished children in 2016 declining from 2015, a possible contradiction. This is odd since WHO and the FAO are both United Nation's agencies and presumably share data. Figure 14 shows the WHO estimates of the number of stunted and underweight children under 5.

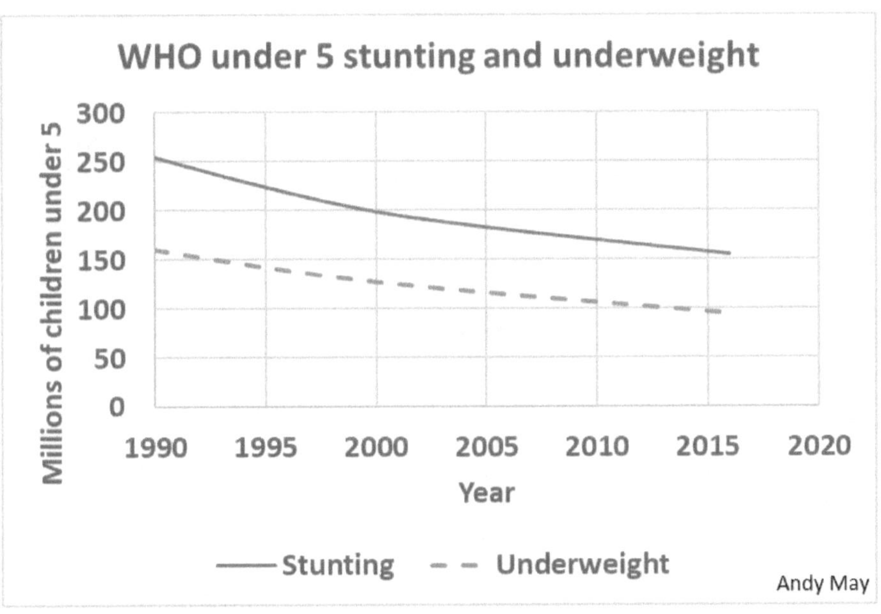

Figure 14. Stunted and underweight children under 5, data source WHO data repository (WHO 2017c).

If the FAO estimate of an increase in total world malnourished is correct and WHO is also correct about the number of malnourished children, the reason is probably an increase in conflict and war. Oddly, many of the same areas with large numbers of undernourished people also have large and increasing numbers of overweight people. Of the

815 million chronically food-insecure and malnourished people, over half, 489 million, live in countries at war (FAO 2017b).

This suggests that the problem is not available food, lowered crop yields or climate change; but production, transportation and delivery problems due to conflict and war. The frequency of wars reached a recent low in 2005 but has increased dramatically since 2010 and is now at an all-time high.

Figure 15 plots the number of conflicts, where at least one participant is a state government, by year. The data are from the <u>Uppsala Conflict database</u> (Allansson, Melander and Themnér 2017) (Gleditsch, et al. 2002) (Upsalla Conflict database 2017).

Figure 15. Worldwide conflicts where a state government is one participant, data source (Upsalla Conflict database 2017).

According to Bjorn Lomborg (B. Lomborg 2007), the number of malnourished is projected to decrease to 108 million by 2080, and if global warming has a negative impact, as it does in some pessimistic forecasts, there would be 136 million malnourished in 2080. Obviously, if wars and conflict continue to rise, the number of malnourished will rise. The point is, how many malnourished people we have is not very dependent upon climate. It is mostly a matter of poverty, war and transportation of goods, especially food. This is well-known in the developing world, but not as well known in the orderly and secure developed world.

Food production

World agricultural production has more than tripled since 1961 (Figure 16), and the increase is more than 4 times in developing countries (Figure 17). According to the U.N., we produce 23 percent more food per person than in 1960, as shown in Figure 18 (FAO

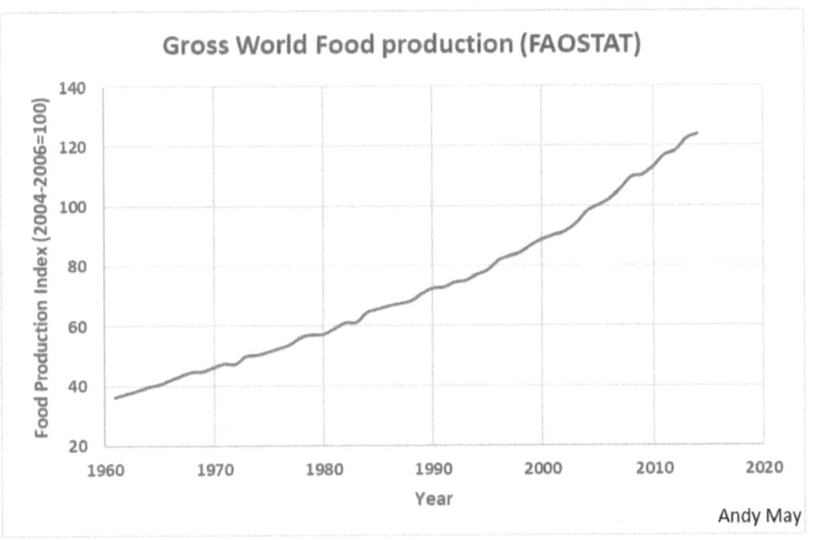

Figure 16. World Agricultural Production by year, data Source (FAO Statistics Division 2017).

Statistics Division 2017). The FAO has <u>recently announced</u> that cereal production reached a new record level in 2017 (FAO 2017c).

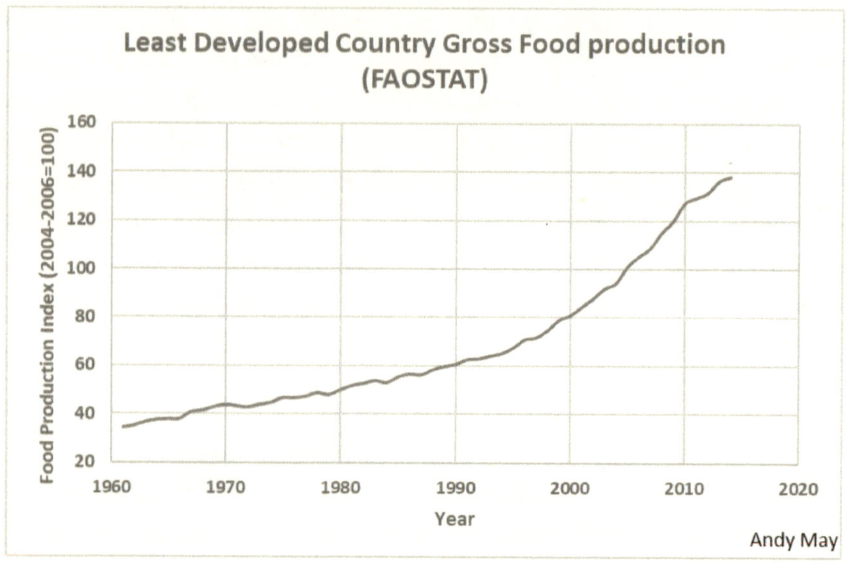

Figure 17. Least developed country agricultural production by year, data source (FAO Statistics Division 2017).

Farming technology, especially fossil-fueled irrigation systems, have dramatically increased the amount of arable land and yields. New, more pest resistant crops and better fertilizers made from natural gas using the Haber-Bosch process have also contributed to the increase in the food supply. This process was created by two brilliant Germans in the early 20th century and has helped feed billions of people. Nitrogen is as essential to life as oxygen or CO_2. In many areas a lack of "fixed" nitrogen is the limiting factor to plant growth. Nitrogen is the most abundant gas in our air, but plants can't use nitrogen in that form; they require nitrogen from ammonia, dung or rotting plants: fixed nitrogen.

Dung and rotting plants are limited, and this reduces crop yields. Carl Bosch and Fritz Haber won the Nobel Prize for their invention of the process to create fixed nitrogen fertilizer (ammonia) from natural gas and air. Fritz Haber discovered the process, and Carl Bosch scaled the process to the factory level with his brilliant engineering skill. Unfortuntely, the Haber-Bosch process was also used to manufacture explosives for Germany in both world wars, but that is another story (Hager 2008).

Food production in the least developed countries is increasing more rapidly than in developed countries because they are adopting western farming technology very rapidly. Besides the adoption of modern farming techniques, human emissions of CO_2 from burning fossil fuels and a warming planet have helped.

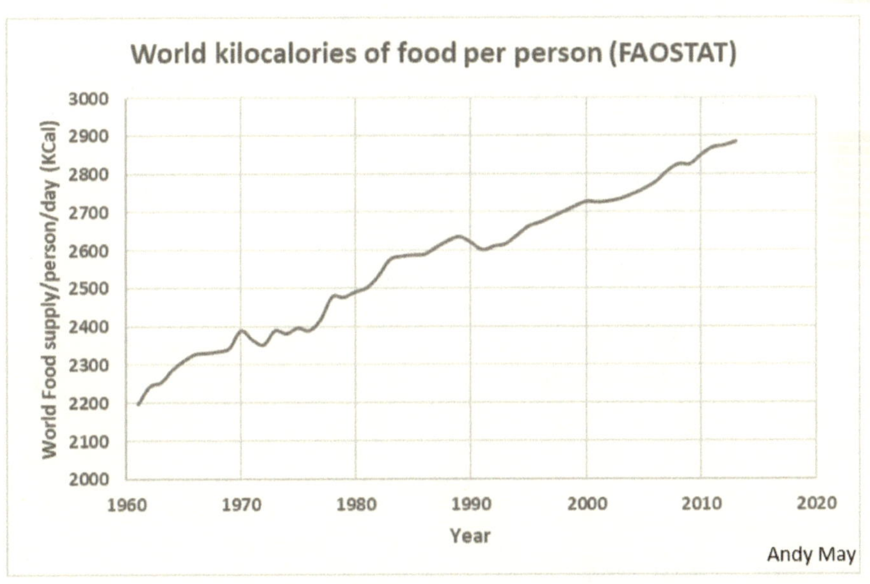

Figure 18. Kilocalories (food Calories) of food per person around the world, data source (FAO Statistics Division 2017).

45

As food production increases around the world, the number of Calories (technically kilocalories) available to each person (Figure 18) is rapidly increasing. Even in Africa and Asia, the available food supply has increased dramatically since 1961.

"Each time our investment in climate saves one person from hunger, a similar investment in direct hunger policies could save more than five thousand people." Lomborg, Bjorn. *Cool It* (Kindle Locations 1618-1619).

I will leave it to the reader to reconcile the data and graphs above with this statement from the IPCC WGII AR5 Technical Summary on page 47 (Field, et al. 2014):

"Based on many studies covering a wide range of regions and crops, negative impacts of climate change on crop yields have been more common than positive impacts (high confidence)."

The data presented in Figures 16 and 17 suggest that any negative effect of climate on crop yields is insignificant. It seems very likely that crop yields will continue to increase.

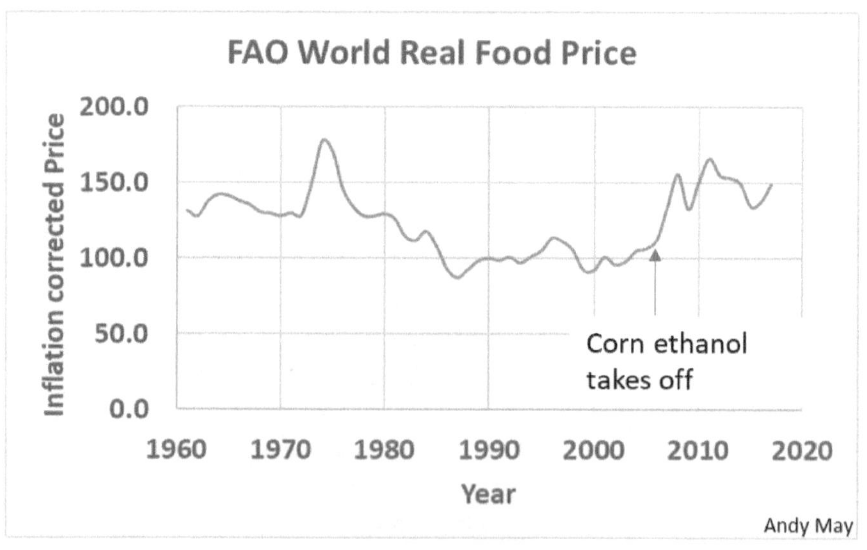

Figure 19. FAO Inflation corrected world food prices (FAO 2017c).

46

World food prices (Figure 19) were flat from 1985 to 2005, and then they increased rapidly when biofuels, especially corn ethanol, took off (see Figure 20) (Joyce 2014). The production of ethanol consumes 40 percent of U.S. corn (Rice 2011). Jordan Schwartz, the lead World Bank economist, discusses the reasons for higher food prices on his blog (Schwartz 2011). He gives three main reasons: speculation in commodity markets, the booming demand in Asia for feed grains and land-use shifting from food quality grains (mainly sweet corn) to biofuel quality corn (field corn) and other biofuel crops. Schwartz wrote his article just before the food price peak in 2012, which corresponds with the flattening of corn-based ethanol production, as seen in Figure 20.

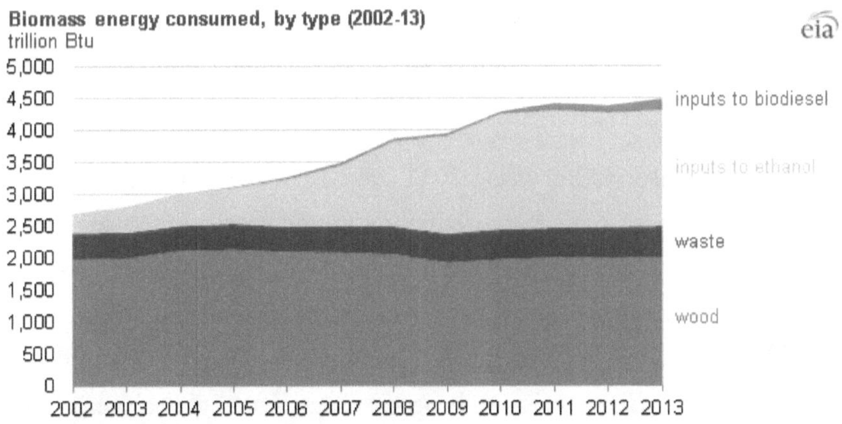

Figure 20. U.S. Biofuel growth (Source: (Joyce 2014)), used with permission.

According to the EIA, corn is the feedstock for nearly all ethanol used in the U.S. (Joyce 2014). The U.S. EPA (Fuelfix 2016) sets the total amount of ethanol that U.S. refiners must purchase. This is problematic because it is <u>unsafe for most gasoline engines</u> (Strauss 2012) to consume fuel that has more than 10 percent ethanol ("E10"). Since the total ethanol mandated by the EPA keeps rising and the amount of gasoline that is being used in the U.S. is not increasing, this creates a "blend wall" (Institute for Energy Research 2016). The blend wall is when refineries are required to buy more ethanol than they can safely put into gasoline. When refineries cannot purchase the mandated ethanol, the refinery is fined. This is one of the rare, and most unfair, situations where a customer is penalized for not purchasing something they cannot use.

One of the reasons that ethanol is added to gasoline is to reduce air pollution. The most common blend is E10, which causes gasoline to burn more completely. However, the net effect of E10 on the environment may not be positive, as discussed by Robert Niven in an article in the journal, *Renewable and Sustainable Energy Systems,* in 2005 (Niven 2005).

Food prices have fallen since 2011, when they reached their peak. Biofuel production growth flattened in 2010, and farms increased their food production to meet demand. The increase in food production and decreasing food prices are also due to improving agricultural technology in the developing world and <u>additional atmospheric CO_2</u> (C. Idso 2011). Still another reason for increased crop productivity is global warming, especially in Siberia and Canada, where it has opened more land to farming (see the IPCC TAR WG2 <u>report</u> on land use changes (IPCC 2001).

Northern Hemisphere warming has contributed to the <u>boom in agriculture</u> (Von der Bouchard and Surk 2016) in Russia. Russia now produces more wheat than the U.S. The Russian boom is partly due to a government effort to stimulate food production, but good weather (meaning warmer weather) also played a large role. Most of recent "global" warming has occurred in western Canada and the U.S., Europe, the Middle East and in Siberia, and has helped farmers in those areas (see Figure 21). The map is made from a grid of annual average temperatures and compares the period from 1951 to 1970 to the period from 2011 to 2017. The period from 2011 to 2017 was chosen so that it includes the 2016 El Niño and the previous cooler period. Most of the "global" warming is in the Northern Hemisphere and over land. There has been little warming in the Southern Ocean and over much of Antarctica.

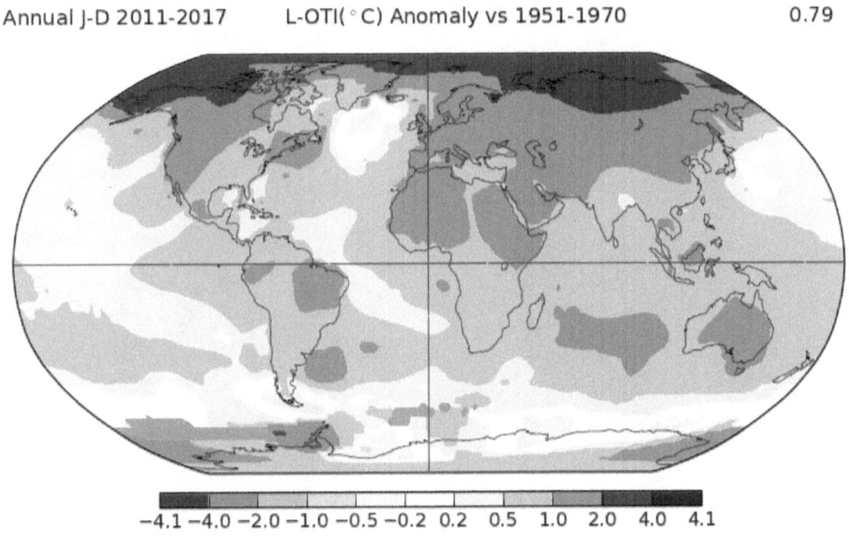

Figure 21. Map of annual global warming intensity in °C. The change from the average from 1951-1970 to the average from 2011-2017, source NASA (NASA 2017b).

Lomborg wrote the following in 1997, and it was true then. What he didn't know at the time was that biofuels would explode onto the scene and humans would foolishly burn their food as fuel by heavily subsidizing corn and sugar ethanol. This had the obvious effect of raising food prices around the world and delaying the reduction of food poverty.

"At the same time as the Earth accommodates ever more people, who are making demands for ever more food, food prices have fallen dramatically. In 2000 food cost less than a third of its price in 1957. This fall in food prices has been vital for many people in the developing world, especially the many poverty-stricken city dwellers. … The fall in the price of food is a genuine long-term tendency. The price of wheat has had a downwards trend ever since 1800, and wheat is now more than ten times cheaper than the price charged throughout the previous 500 years." Lomborg, Bjørn. *The Skeptical Environmentalist: Measuring the Real State of the World* (p. 62).

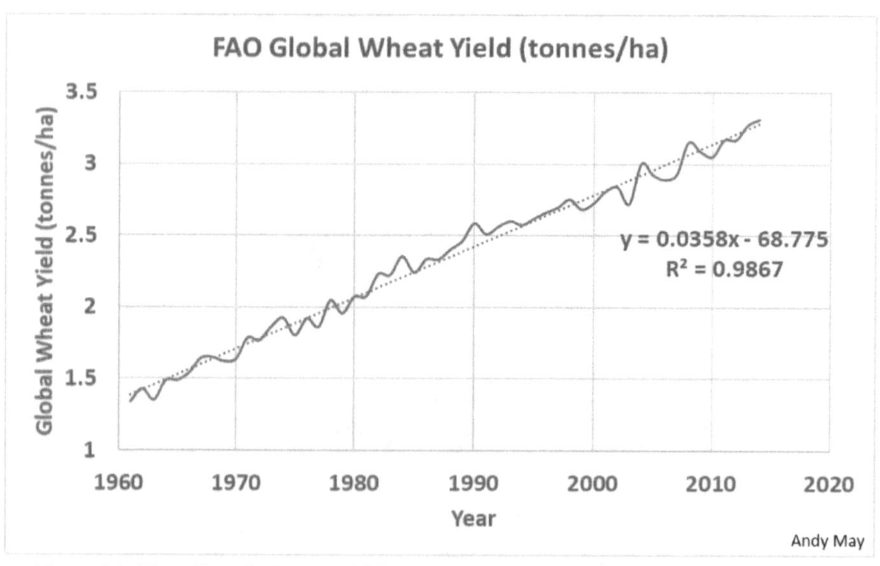

Figure 22. Plot of total wheat yield from 146 countries, data from Ourworldindata.org (Roser and Ritchie 2018) and the FAO (FAO Statistics Division 2017).

The AR5 WGII report rather inexplicably reports that the impact of climate change reduced agricultural yields from 1960 to 2013 (see Figure TS.2E, page 43) (Field, et al. 2014). This is at odds with actual yield data from the United Nations FAO (Figure 22). The ourworldindata.org (Roser and Ritchie 2018) web site plots FAO crop yield data that shows yields increasing all over the world, not falling. Figure 22, as an example, plots total wheat yield from 146 countries from 1961 through 2014, using 2017 FAO data.

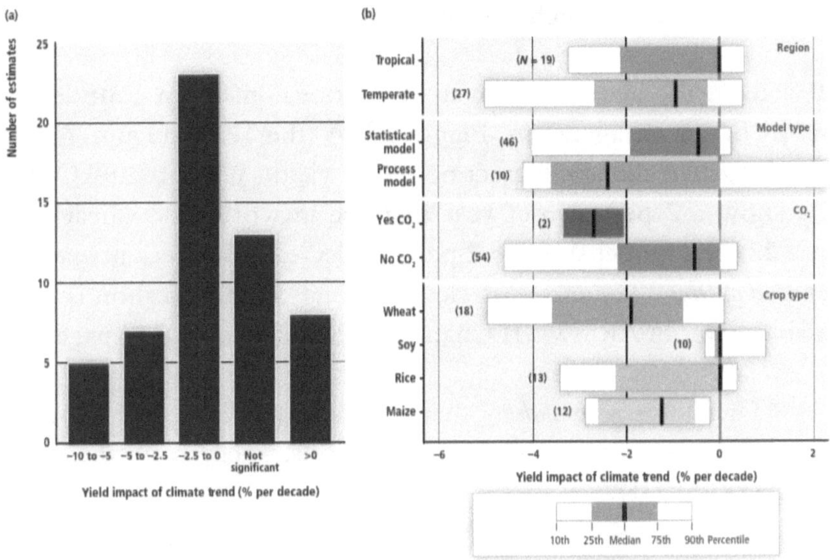

Figure 23. Plots of studies of the impact of climate change on crop yields, source IPCC AR5 WGII, Chapter 7, page 492 (Porter, et al. 2014). Used with permission.

Global wheat yields are increasing at a rate of over 2 percent per year (0.04 tonnes/ha/year) from 1961 to 2014, and the observed increase is very linear. Total food production is also increasing, and the rate of increase in food production is increasing, as can be seen in Figures 16 and 17. The IPCC AR5 WGII conclusions are model-based and show very tiny "negative climate caused" changes of up to 2 percent/decade (see Figure 23) this figure is from Chapter 7, page 492 of the full report (Porter, et al. 2014).

The IPCC model was run for the period 1960 to 2013. AR5 WG2 explains their model results as follows, Chapter 7, page 491 (Porter, et al. 2014):

> "Based on these studies, there is medium confidence that climate trends have negatively affected wheat and maize production for many regions (Figure 7-2 [our Figure 23]) (medium evidence, high agreement)."

Based on the actual data over the period, "medium confidence" seems to be an exaggeration. Figure 23(b) (the IPCC Figure 7-2(b)) shows a negative climatic impact on wheat yield. While the FAO data clearly show a 2 percent per year increase in worldwide wheat yield (Figure 22), the model shows a 2 percent decrease per decade over the *same period of time*. The model is clearly wrong, so the question is why? We also see this in AR5 WGII Chapter 7 (Porter, et al. 2014), page 491:

> "There is *high confidence* that warming has benefited crop production in some high latitude regions, such as northeast China or the UK."

The apparent contradiction between this statement and Figure 23(b) is not explained in the text, but Figure 23(a) does include some positive estimates of climate impact. The vertical scale of Figure 23a indicates the number of studies with that estimate, and the numbers in parenthesis in Figure 23b are also the number of studies displayed in the box plot. The caption for the IPCC figure says that some of the studies summarized in our Figure 23 included effects of positive carbon dioxide trends, but most did not (page 492, AR5 WG2). Figure 23b suggests only two studies included the fertilization effect of CO_2.

Craig Idso and Sherwood Idso, in their web site CO2science.org (C. Idso, CO2Science.org 2017b), identify many papers that show increased growth of corn and sorghum due to increased CO_2. Three of the papers are (Maraco, Edwards and Ku 1999), (Ottman, et al. 2001), and (Watling and Press 1997). There are also many papers on the benefits of additional CO_2 on wheat and other cereals, such as (Manderscheid and Weigel 2007), (Veisz, et al. 2008), and (Robredo, et al. 2007). Many more papers on the beneficial effects of additional CO_2 on crops listed in Figure 23b are cited <u>in</u> (C. Idso 2017c). Elevated CO_2 makes crops more drought resistant and decreases their use of water per pound of growth. Clearly, Figure 23b is not a fair presentation of the available peer-reviewed literature regarding the effects of additional CO_2 on crops. It is a highly selective sample.

Much of the IPCC AR5 discussion on food supply relies on data and analysis by (Lobell, Schlenker and Costa-Roberts 2011). Lobell is also one of the authors of the chapter. This paper relies on a comparison of grain yields from 1960 to 1980 to those from 1980 to 2008. The computer model used only considers changes in temperature and precipitation and ignores adaptation or new technology. From the paper:

"Any model has its limitations, and we recognize a few caveats that are common to statistical models. Our approach may be overly pessimistic because it does not fully incorporate long-term adaptations that may occur once farmers adjust their expectations of future climate. Examples of this would include expansion of crop area into cooler regions, switches to new varieties, or shifts toward earlier planting dates, although there is little evidence that the latter is happening beyond what is expected from historical responses to warm years." (Lobell, Schlenker and Costa-Roberts 2011)

Further, both Lobell, et al. and the IPCC AR5 WGII Chapter 7 spend many pages discussing their "modeled results" and not one word on the fact that yields are, in fact, increasing at a rate of 2 percent per year. Lobell, et al. mention "stagnation of yields" in France and "depressed" corn and wheat yields in India. FAO statistics for these countries, as well as China and Argentina, are shown in Figure 24. These figures can be compared to Figure 22, which is the average wheat yield for 146 countries. Not shown are plots for Canada, Russia and the U.S., which all show a steady increase in wheat yields from 1990 to 2014. The data suggests that the "stagnation" and "depressed" local yields in France and India are anomalies and not due to any global climatic effects.

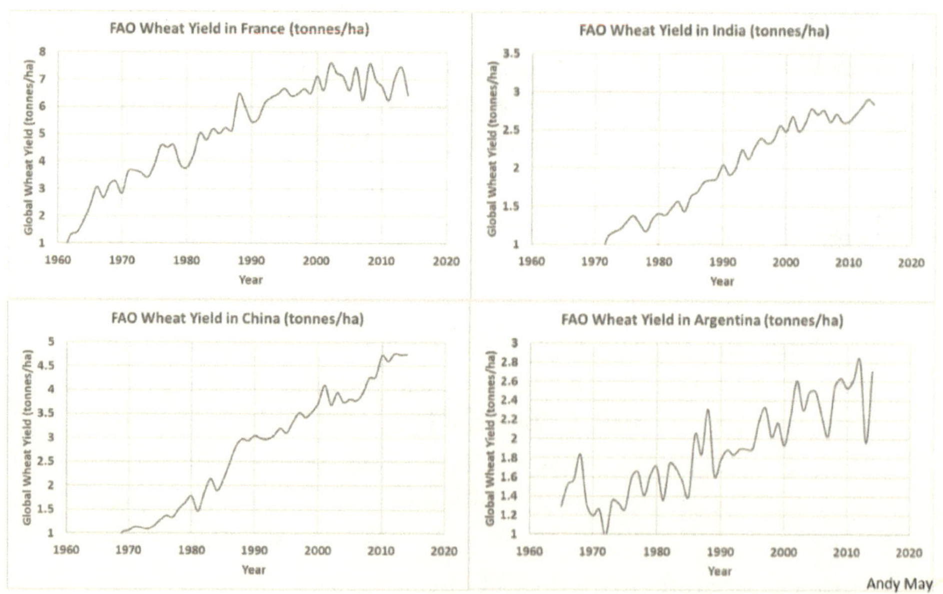

Figure 24. Wheat yields in four countries, data source Ourworldindata.org (Roser and Ritchie 2018) and the FAO (FAO 2017b).

Chapter 2 Conclusions

A Mathusian or Erhlichian food catastrophe seems highly unlikely in the foreseeable future. Food production is growing rapidly due to the spread of western farm technology, CO_2 fertilization and new arable land in Canada and Russia due to global warming. The additional food is being distributed worldwide as the number of kilocalories (Calories) of food available to people worldwide is increasing dramatically.

Although IPCC models show that climate change should be decreasing crop yields, the data show no slowing of growth in yields. If it is in the models but not in the data, it seems likely that the models are wrong and there is nothing to worry about.

Although the world population continues to increase, the number of people being added to the population each year is decreasing. In addition, the rate of population growth has been decreasing since 1963. The peak growth year in absolute numbers was 88,000,000 in 1989, and it is still decreasing. Both numbers are expected to be near zero in 2100. The growth increased due to an increasing average life expectancy, which was due to a falling death rate. The birth rate then began to decline as the population became more urban and children became less valuable as workers and for security in old age. That is, the cost of raising a child increased, and the benefits of having children decreased. As the world becomes more prosperous and urban this trend will continue. Human prosperity is the key to a better world.

Chapter 2 Summary

While the IPCC and others assert that climate change is projected to negatively impact agricultural food production, no slowdown in the growth of crop yields or total food production is seen in the United Nations Food and Agricultural Organization (FAO) data. Further the growth in the world food supply is accelerating, especially in less developed countries, and the total average food available to people is also growing. Projections, unsupported in the data, should not be used to make policy or to draw conclusions.

While the world population is growing, both the rate of growth and the number of people added each year are decreasing. The food supply is growing as more farmers adopt western farming technology. The key components of higher crop yields are fossil-fueled irrigation systems, additional fixed nitrogen fertilizer from natural gas and the Haber-Bosch process, and additional CO_2 from burning fossil fuels.

To benefit mankind and the environment, we should strive to make the world more prosperous. Prosperity improves both human well-being and the planet. Reducing or eliminating fossil fuels, nuclear power or hydroelectric power does the opposite.

In the first two chapters we have tried to deal with the more extreme "Climate-change-will-end-the-world-and-kill-us-all!" claims. In Chapter 3 and subsequent chapters, we will discuss the cost of trying to mitigate warming by curtailing or ending fossil fuel use with global programs like the Kyoto or Paris agreements versus the cost of each community dealing directly with any local climate change problems that are affecting them. Every community has its own problems: those on the ocean may have to deal with rising sea level, high in the mountains it may be melting glaciers, and in other areas flooding or drought may be a problem. Mitigation is not the only possible solution to climate change. Adaptation is another that should be considered as an alternative. As we will see, mitigation may not be an effective solution in any case.

CHAPTER 2

CHAPTER 3

The Cost of Global Warming

Hopefully, the first two chapters have convinced the reader that man-made climate change and global warming are not an existential threat to humanity or the planet. This leaves us in a discussion about the cost of global warming, which is something we can calculate. To do the calculation, we need to estimate the monetary damages caused by global warming, when they will be incurred, and the discount rate of money over time. We will not attempt the calculation here, as it is too complex, but we will discuss the parameters and some calculations done by others.

We should remind the reader that the assertion that humans have caused most of the recent global warming has not been proven. The calculation of human influence on climate is based only on unvalidated climate models, as <u>discussed</u> by (Curry 2017). In fact, most climate models <u>cannot model the global warming</u> (May 2016c) from 1910 to 1945. If they can't hindcast known global warming, how can they accurately forecast it? <u>Forecasting natural warming</u> (Meehl, et al. 2016) is obviously critical to computing the magnitude of man's impact, so we must remain skeptical of any calculation of man's influence. This is discussed in more detail in (May 2016e).

To compute a cost for global warming, we first assume it has a cost. Then we must assume a value, in lives and treasure, for the cost. We also need to assume a timetable of the costs and benefits so we can apply a discount rate for the money spent and the money saved. One might say dubious climate model results are fed into dubious economic models, and projections are then made for 100 years.

So, we will assume that human-caused global warming is a danger. We only make this assumption to show that eliminating fossil fuels is not necessarily the correct choice even if the dangers exist. The IPCC WGII AR5 states this with a little more certainty than they should (page 37 of the technical summary):

"Human interference with the climate system is occurring. Climate change poses risks for human and natural systems." (Field, et al. 2014)

Humans certainly affect the climate to some degree, as many major species do, especially trees and phytoplankton. Climate change, whether natural or man-made, is a risk. There are always two options in dealing with climate changes: we can adapt to the changes or, if we are the cause, we can mitigate them by changing our behavior. The two options can conflict. If mitigation reduces our use of fossil fuels, a cheap form of energy, and we adopt an alternative fuel that is more expensive, we reduce our ability to adapt.

Adaptation can involve constructing seawalls, levees and dikes, installing air conditioners, irrigating dry farm land, and many other things that require a lot of energy and transporting large quantities of materials and fuels. Global warming risks, adaptation and mitigation methods are described in the <u>IPCC WGII AR5 technical summary</u> (Field, et al. 2014).

What is the ACTUAL cost of global warming?

Estimating future costs due to global warming is exceedingly difficult. From the IPCC WGII AR5 Technical Summary, page 71:

"Global economic impacts from climate change are difficult to estimate. Economic impact estimates completed over the past 20 years vary in their coverage of subsets of economic sectors and depend on a large number of assumptions, many of which are disputable, and many estimates do not account for catastrophic changes, tipping points, and many other factors. With these recognized limitations, the incomplete estimates of global annual economic losses for additional temperature increases of ~2°C are between 0.2 percent and 2.0 percent of income (±1 standard deviation around the mean) (medium evidence, medium agreement)."

World GDP is currently about $75.5 trillion dollars per year according to the <u>World Bank</u> (World Bank 2017b), so 1 percent is about $0.755 trillion, which is a lot of money, but that is the IPCC "medium evidence and medium agreement" estimate of the damage global warming can do, ± about *100 percent*. That is, from 0.15 trillion to 1.51 trillion dollars by 2100!

The IPCC estimate of the cost of global warming is not accurate by any means. The others I will show in this chapter are not either. But, the estimates are used by pundits, ex-Vice-Presidents, and high school dropout actors, etc. all the time to try and scare us. Let's examine them.

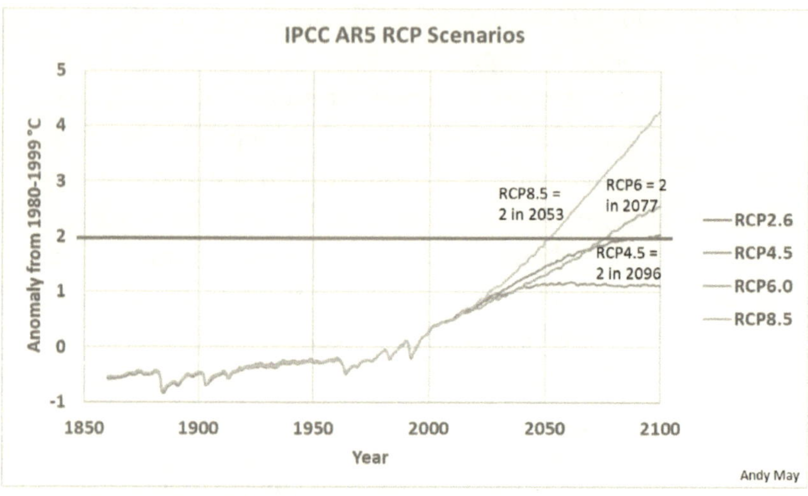

Figure 25. IPCC AR5 surface temperature projections (data source: IPCC AR5, data downloaded from the KNMI climate explorer).

The quote is specific about the "dangerous temperature" of ~2°C but does not give us a maximum "safe" temperature. In Figure 25 we plot four of the IPCC AR5 model ensemble mean surface temperature predictions. There is no data on the plot, just predictions and hindcasts.

The RCP ("Representative Concentration Pathway") is a name given to a set of climate model parameters. The parameters include predicted greenhouse gas emissions and assumptions about how sensitive the climate is to those emissions. In this case the RCP2.6 scenario is a "safe" set of assumptions, so we assume that the warming under this scenario will cause no damage. This sets a limit of 1.1°C over the 1980-1999 mean temperature. RCP8.5 is a fairly radical "worse case" scenario and considered <u>unlikely</u> (Ritchie and Dowlatabadi 2017). Under this set of parameters 2°C is reached in 2053. For the sake of argument, we will focus on the RCP4.5 scenario as most analysts do. In this scenario the "dangerous" 2°C temperature is reached in 2096.

Mitigation and adaptation

Since the IPCC is convinced that man-made CO_2 is the main driver of climate change, they believe that mitigation must be a component of the climate change solution. The IPCC AR5 WGIII technical summary describes mitigation on page 37 (Edenhofer, et al. 2014):

"'Mitigation', in the context of climate change, is a human intervention to reduce the sources or enhance the sinks of greenhouse gases (GHGs)."

The IPCC AR5 WGII technical summary describes adaptation in this way on page 40 (Field, et al. 2014):

"The process of adjustment to actual or expected climate and its effects. In human systems, adaptation seeks to moderate or avoid harm or exploit beneficial opportunities. In some natural systems, human intervention may facilitate adjustment to expected climate and its effects."

We adopt these definitions for this chapter.

In any discussion of potentially dangerous global warming, several potential costs or hazards are bound to be brought up. Aren't hurricanes and other severe weather events more severe due to climate change? We will discuss this in Chapter 7. Isn't sea-level rising at a dangerous and accelerating rate? This is the subject of Chapter 8. Isn't the cost of global warming going to be astronomical? The topic of this chapter. Won't there be far more heat-related deaths? The subject of Chapter 6. There are peer reviewed papers, cited in <u>AR5</u> (IPCC 2014), that predict these scenarios. The predictions are based on unvalidated models, but there are little or no actual data to support any of them. We will compare the predictions to the data. Many of the benefits of additional atmospheric CO_2 were discussed in Chapter 2. We will not cover that ground again.

The recent rise in global average temperature, whether natural or man-made, is not unusual, as we will discuss in Chapter 4. But, for the sake of argument, if future warming might be harmful and if man-made CO_2 is causing most of the current warming, is it better to reduce fossil fuel use? Or should we spend our money adapting to the warmer temperatures? To answer these questions, we must be quantitative. Reducing fossil fuel use increases the cost of energy (Weißbach, et al. 2013) and reduces our ability to adapt, since all adaptation methods require energy. From AR5 WGIII, Technical Summary, page 67 (Edenhofer, et al. 2014):

> "Particular mitigation actions can affect sectoral climate vulnerability, both by influencing exposure to impacts and by altering the capacity to adapt to them."

The time value of money and the discount rate

Reducing fossil fuel use has an immediate impact on our well-being, for a possible and unproven improvement on well-being over 100 years from now. It is fairly likely that currently we are benefiting from the additional CO_2 in the atmosphere, and it will <u>continue to be beneficial</u> (May 2016d) for at least several more decades. NASA researchers have described how CO_2 is making the Earth <u>greener</u> (Reiny 2016).

From the IPCC AR5 WGIII technical summary, page 61 (Edenhofer, et al. 2014):

"Investments aimed at mitigating climate change will bear fruit far in the future, much of it more than 100 years from now. To decide whether a particular investment is worthwhile, its future benefits need to be weighed against its present costs. In doing this, economists do not normally take a quantity of commodities at one time as equal in value to the same quantity of the same commodities at a different time. They normally give less value to later commodities than to earlier ones. They 'discount' later commodities, that is to say. The rate at which the weight given to future goods diminishes through time is known as the 'discount rate' on commodities."

In other words, the "present value" of benefits paid in the far future is very small, due to the "time value of money." Or, one could simply say: "A bird in the hand is worth more than two birds in a bush." If we assume climate change is dangerous and that man's emissions are the cause, two big "ifs," then how much do we spend today to avert dangerous climate change in 100 years or more? If the dangers were certain and known, some expenditure would be worth it. But, if the prospective dangers are like those predicted by Thomas Malthus in 1798 or Paul Ehrlich in 1968 (see Chapter 2), not so much. If we wait until we are sure of the dangers, is the wait dangerous? These are the questions that we need to address.

The <u>net present value</u> calculation that I allude to above is very sensitive to the choice of a discount rate. This is particularly true of investments in climate change mitigation or adaptation because the payout is 100 years in the future or longer. Because the U.S. S&P 500 collection of stocks, between 1975 and 2017, has an average annual <u>rate of return of 7.5 percent</u> (PK 2017), adjusted for inflation, with dividends reinvested, this value is reasonable to use as a discount rate.

The idea is that if you calculate a positive net present value at 7.5 percent you beat the stock market rate of return with your investment. But, if you can't beat the S&P 500, just put your money into the stock market instead. Discount rates between 6 percent and 12 percent can be used and justified. Before 2100, when all our climate change money goes out and none is coming in, 6 percent loses twice as much money as 12 percent, but both lose a lot of money. This money comes out of taxpayers' pockets and goes to alternative fuel manufacturers or other organizations meant to reduce greenhouse gas emissions.

Dr. Bjorn Lomborg, Dr. Richard Tol, Alex Epstein, Dr. Matt Ridley, and Dr. Roger Pielke Jr. have written extensively on the economics of climate change. This chapter is mostly based upon their work and the IPCC AR5 report. We have especially relied upon Lomborg's *Cool It*, Epstein's *The Moral Case for Fossil Fuels*, Ridley's *The Rational Optimist*, and Pielke Jr.'s *The Climate Fix*.

For the most part, they do not question the idea that man is causing some of the recent global warming; they simply use the IPCC climate forecasts to determine the costs of global warming to society. While it has not been proven that man is the cause of most of the warming since the Little Ice Age (May 2016e), it is still useful to discuss the possible costs of warming since it ended around 1850 and various proposed solutions. The Little Ice Age was the coldest period in the Holocene. It is not a climate we should be trying to return to. This is discussed in more detail Chapter 4.

We list some commonly claimed costs (or hazards) of global warming and, when appropriate, the potential costs of either adapting to the warming or mitigating the temperature change through reductions in fossil fuel use. We will only discuss a few of the more well-known claimed costs, because to address all of them would be impossible. Climate change has been blamed for a large number of problems, including stunting <u>Chinese manufacturing</u> (Zhang, et al. 2018), starting the <u>Syrian civil war</u> (Selby, et al. 2017), decimating <u>bumble bees</u> (Baron, et al. 2016), and so on. A comprehensive list of claimed costs and hazards can be seen in the AR5 WGII technical summary. We will stick to the more conventional claims.

The cost of reducing CO$_2$ emissions

There have been two international agreements to mitigate climate change by reducing fossil fuel use. They are the <u>Kyoto</u> (United Nations Climate Change 2013) and <u>Paris</u> (United Nations Climate Change 2018) agreements. These agreements assume that current climate models are correct, and temperatures are mostly increasing due to human fossil fuel emissions. The countries in the agreements pledged to reduce their fossil fuel emissions to mitigate global warming. Here we discuss the cost of this mitigation to achieve a baseline cost.

The Kyoto treaty, if it had been implemented, was expected to cost \$180B/year, or 0.5 percent of global GDP, from (B. Lomborg 2007), original source (Weyant and Hill 1999).

> "For the full Kyoto Protocol with the United States participating, the total cost over the coming century turns out to be more than \$5 trillion. There is an environmental benefit from the slightly lower temperature toward the end of the century: about 0.3°F. The total benefit for the world comes to almost \$2 trillion." Lomborg, Bjorn. *Cool It* (Kindle Locations 610-613).

Curtailing the fossil fuel industry would have additional repercussions. Since the oil and natural gas industry revenue was about

$5 trillion/year in 2014 (Investopedia 2015), or 5 percent of global GDP, limiting or eliminating this industry would cost jobs and economic disruption. So, the excess cost of mitigation, over the value of the benefits, could be even higher than the estimated $3 trillion noted above, depending upon what happens to the fossil fuel industry.

If reducing or eliminating fossil fuels reduces world GDP, as most expect, it will increase the number of poor and lower our standard of living. Both of these factors will affect world health, as noted in (Pritchett and Summers 1996) a *Journal of Human Resources* article entitled "Wealthier is Healthier."

The Paris Agreement

If the U.S. had stayed in the Paris agreement, it would have committed the U.S. to reducing its carbon dioxide emissions by 26 percent-28 percent, relative to 2005, by 2025. This was not achievable with the regulations currently in place, and additional regulations and/or taxes would have been required. The agreement was not binding, but the pledges the Obama administration made could have cost the U.S. as much as $3 trillion and 6.5 million jobs according to some opponents of the plan (Cama and Henry 2017).

Other estimates of the cost of the additional regulations required, including the now defunct CPP (Clean Power Plan), vary from $37 billion (RFF) (Chen and Hafstead 2016) to $250 billion (NERA) (Bernstein, et al. 2017) in lost GDP by 2025. Economy-wide job loss estimates vary from no job losses (RFF) to more than 2.7 million by 2025 (NERA). The higher GDP and job losses are from a U.S. Chamber of Commerce study by NERA. Lower estimates are from a liberal think tank "Resources for the Future" (RFF) article. Both studies acknowledge that regulations and efforts in place in 2016 were not sufficient to meet the U.S. Paris commitments. The Heritage

<u>Foundation</u> (Dayaratna, Loris and Kreutzer 2016) estimates the U.S. would lose 206,104 manufacturing jobs alone (see Figure 26).

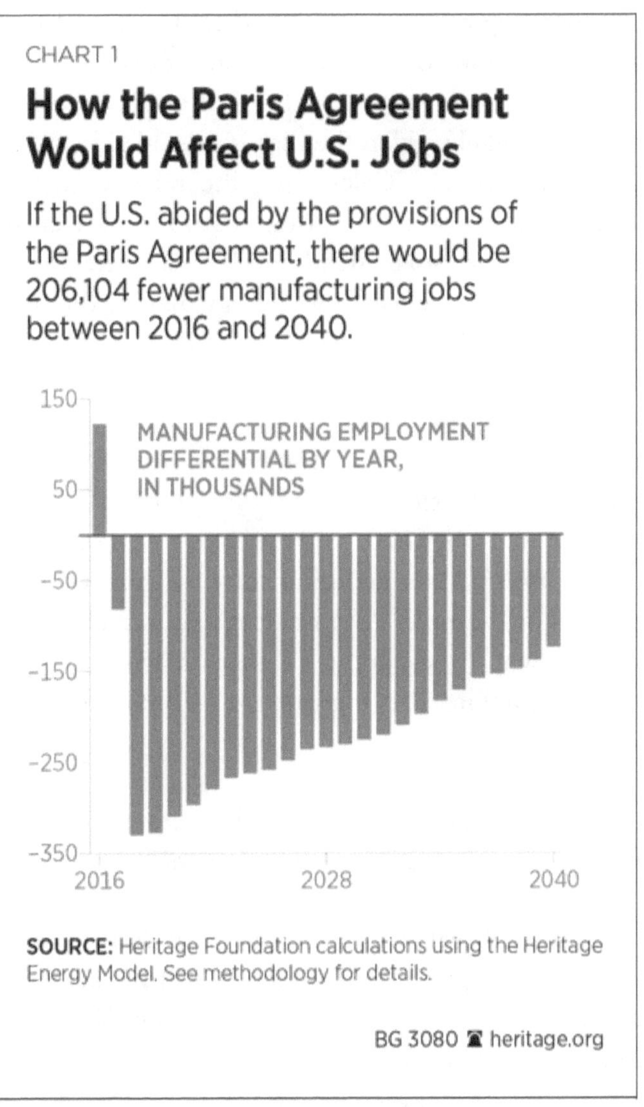

Figure 26. The effect of the Paris Agreement on U.S. jobs, source Heritage.org (Dayaratna, Loris and Kreutzer 2016). Used with permission.

The differences in the estimates stem from differing views on the relative future cost of energy from alternative sources and fossil fuels and the costs of meeting the U.S. emission reduction commitments. The NERA estimate uses current cost estimates for alternative fuels that are generally accepted by the EIA and IEA. The RFF essentially assumes, using projections of the current rate of decline of alternative fuel costs, that alternatives will be as cheap, if not cheaper, than fossil fuels. This is very unlikely, given the reasons detailed in (Weißbach, et al. 2013) and (May 2017c). Calculating the relative cost of energy produced using different sources is quite complex, but currently and for the foreseeable future solar, wind and biomass energy sources are net negative. That means we use more energy making the fuels and/or operating the power plants than we get out of them. For the details of the calculations involved, see (Weißbach, et al. 2013).

Computing job losses depends upon how expensive western energy becomes as they curtail fossil fuel use. India and China will be allowed to use all the fossil fuels they like, so their energy costs will be very low compared to the West. Manufacturing jobs will move to where the cheap energy is. For example, cheap natural gas and electricity are the principle reasons why <u>Voestalpine</u> (Tirone 2013) and <u>Tanjin Pipe</u> (TPCO 2015) (Welitzkin 2016) group moved facilities to Corpus Christi, Texas.

Due to higher paying manufacturing job losses, people will migrate to lower paying service sector jobs, which lowers family incomes.

Absent a dramatic breakthrough in fusion or conventional nuclear power, renewables, except for hydroelectric, will remain prohibitively expensive unless heavily subsidized. Conventional nuclear is economic, but regulations make it very difficult to permit and install. Fusion and thorium reactors are not commercialized yet, but if they are they could be a game changer.

The NERA assumptions are more realistic. There are no credible scenarios where alternative fuels could be as cheap or cheaper than fossil fuels without an enormous carbon tax that would drastically lower our standard of living. This is a policy very unlikely to succeed or be very popular in the developed world, especially in the U.S.

The _New York Times_ (Popovich 2017) likes to say that more people work in the solar power generation industry than in the coal power generation industry. This sounds great until you realize that in 2016, according to the EIA (EIA 2017), solar produced 0.9 percent of our electricity with 373,807 people [from a U.S. Energy Department study (U.S. Energy Department 2017)] versus coal, which produced 30 percent of our electricity with 160,119 people from the same Energy Department study.

Productivity growth is where our standard of living comes from. Switching to solar is going in the wrong direction. It is apparent that the _New York Times_ is ignorant of basic economic principles. What this means is that it takes 83 solar workers to produce the same amount of electricity as one coal worker. This only makes economic sense if the solar workers are paid 1.2 percent of a coal worker's pay.

To achieve the U.S. Paris commitment the RFF recommends imposing a $21.22 (in 2013 U.S. dollars) per ton tax on CO_2 emissions. This is a crippling tax given that the best estimate of the costs associated with CO_2 emissions is $2 per ton and the costs of the warming are almost certainly less than $14. We would be paying $7 to almost $20 per ton more than the costs we are avoiding. That is guaranteed to lower our standard of living. From _Cool It_ (B. Lomborg 2007):

"In a global macroeconomic model, the total present-day cost for a permanent one-dollar CO_2 tax is estimated at more than \$11 billion. So, we might want to think twice about cranking up the knob to a thirty-dollar CO_2 tax, which will cost almost \$7 trillion. ... [Richard Tol] finds that with reasonable assumptions the cost is very unlikely to be higher than fourteen dollars per ton of CO_2 and likely to be much smaller. When I specifically asked him for his best guess, he wasn't too enthusiastic about shedding his cautiousness—true researchers invariably are this way—but gave a best estimate of two dollars per ton." Lomborg, Bjorn. *Cool It* (Kindle Locations 558-578).

The Paris agreement would only have eliminated coal-related jobs in the U.S. and other western countries, essentially transferring these jobs to India and China where coal use will continue to increase without penalty. This is also true of the steel industry and some portion of the oil and gas industry. Basically, it was a very bad deal for the U.S., and the U.S. withdrew from the agreement for that reason (Harbin 2017).

Costs versus benefits

World energy consumption continues to increase. Even in the U.S. total energy consumption is increasing (see Figure 27). Appliances, houses and cars are more efficient today, but we use energy, especially electricity, to power many more things these days. Primary energy is the original energy source, such as burning coal or natural gas. Solar panels or wind farms are also primary sources. Table 1 lists the primary energy sources for producing electricity in the U.S. in 2016 (EIA 2017b).

2016 EIA Primary Energy Sources for producing electricity		
	(Trillion Btu)	% of total
Coal	12,996	34.5%
Natural Gas (Excluding Supplemental Gaseous Fuels)	10,301	27.3%
Petroleum (mainly diesel or fuel oil)	244	0.6%
Total Fossil Fuels	23,542	62.4%
Nuclear Electric Power generation	8,427	22.3%
Conventional Hydroelectric Power	2,459	6.5%
Geothermal Energy	146	0.4%
Solar Energy	328	0.9%
Wind Energy	2,094	5.6%
Biomass Energy	505	1.3%
Total Renewable Energy	5,531	14.7%
Electric Power Sector Net Imports	206	0.5%
Total Primary Energy Consumed	37,705	100.0%

Table 2. Primary energy sources for producing electricity in the U.S. in 2016, data source (EIA 2017b).

Fossil fuels are the primary energy source for 62 percent of U.S. electricity. Adding nuclear and hydroelectric to the fossil fuels brings the total to 91 percent. Due to energy losses in the electricity generation process and transmission, only 35 percent (up from 20 percent in 1949) of the primary energy consumed is delivered to the end-user as electricity (EIA 2017b). Electricity is very popular, and we use many more electrically powered appliances today than we did in the past. We would much rather plug our phone into a wall socket than take delivery of a load of coal and turn it into electricity ourselves. But, electricity is a very wasteful way to consume energy, and as the energy delivered as electricity grows to 10 percent, the total amount of primary energy consumed rises 28 percent, a factor of 2.8, as shown in Figure 27. Thus, going electric uses more primary fuel and is not more efficient or environmentally friendly, as many assume.

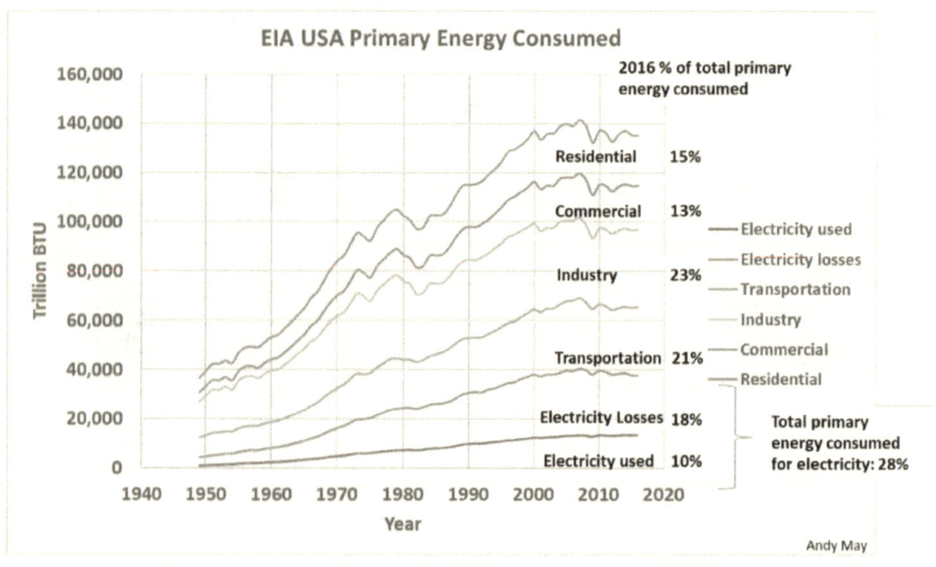

Figure 27. Primary energy (mainly coal, natural gas, petroleum, and renewables) consumed by sector, including the fuel lost making and delivering electricity, data source: (EIA 2017b).

According to the BP Energy Outlook for 2017, total energy consumed in the world has quadrupled since 1965, and it is projected to continue to increase, as shown in Figure 28.

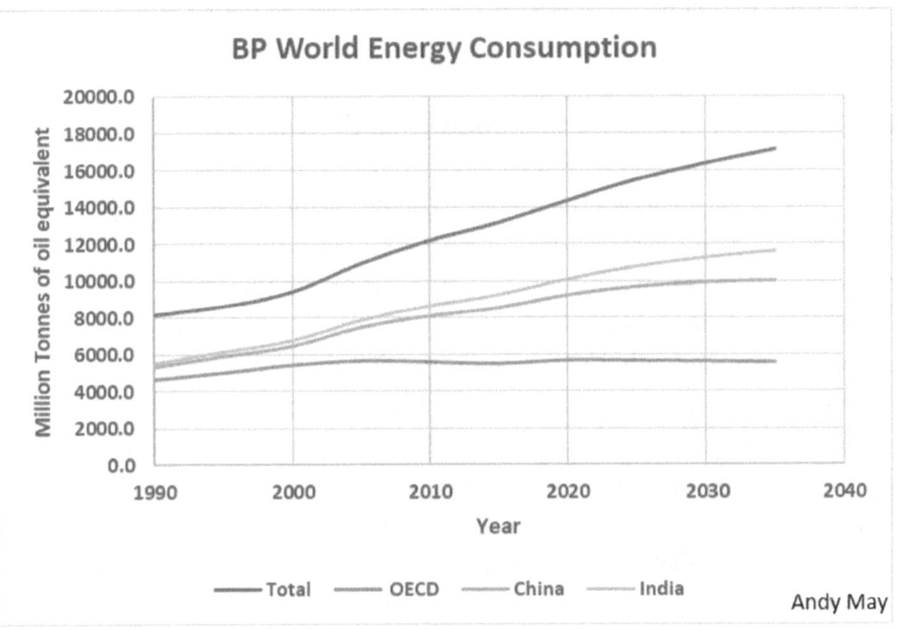

Figure 28. World energy consumption, data source (BP 2017).

Whether we discuss the costs and benefits of Kyoto or Paris, the problem is the same. It is very hard to get the costs of reducing CO_2 emissions down to a level that makes economic sense. Lomborg explains:

> "Stabilizing the temperature increase to 4.5°F … cost[s] $15.8 trillion. It is instructive to compare this to the total cost of global warming. Models show that if global warming wasn't happening, the world would end up about $14.5 trillion richer.

Thus, the cost of global warming can be estimated at $14.5 trillion. Stabilizing the temperature increase at 4.5°F ... means we end up paying more for a partial solution than the cost of the entire problem. That is a bad deal." Lomborg, Bjorn. *Cool It* (Kindle Locations 646-650).

The time value of money is also important in this analysis:

"Thus, as one academic paper [(Kavuncu and Knabb 2005)] points out, 'the costs associated with an emissions stabilization program are relatively large for current generations and continue to increase over the next 100 years. The first generation to actually benefit from the stabilization program is born early during the 24th century.' If our desire is to help the many generations that come before then, along with the world's poor, cutting emissions is not the best way. Perhaps more surprisingly, cutting emissions is also not the best way to help people in the 24th century, since we could have focused on solving many other immediate problems that would leave the far future much better off." Lomborg, Bjorn. *Cool It* (Kindle Locations 672-677).

"Kyoto is an extraordinarily expensive way of doing very little good far into the future," Lomborg, Bjorn. *Cool It* (Kindle Locations 637-638).

Chapter 3 Conclusions

Can climate models be devised to predict that human emissions can cause great financial damage? Yes, they can be constructed, but are they accurate? Are they accurate enough that we must reduce our fossil fuel use? Reducing fossil fuel use raises the cost of energy and makes us poorer, less adaptable, and more vulnerable to climate disasters.

"Alarmism has a long history in the climate debate. Perhaps most chillingly, this was evident in the witch trials in medieval Europe. After the Inquisition's eradication of the actual heretics (like Cathars and Waldensians), most witches from the early 1400s onward were accused of creating bad weather. The pope in 1484 recognized that witches 'have blasted the produce of the earth, the grapes of the vine, the fruits of the trees, … vineyards, orchards, meadows, pasture-lands, corn, wheat and all other cereals.' As Europe descended into the Little Ice Age, more and more areas experienced crop failure, high food prices, and hunger; witches became obvious scapegoats in weakly governed areas. As many as half a million individuals were executed between 1500 and 1700, and there was a strong correlation between low temperatures and high numbers of witchcraft trials across the European continent." Lomborg, Bjorn. *Cool It* (Kindle Locations 1825-1833).

Even when the cause is obviously natural, society still needs to find someone to blame. For more on this topic see (May 2016f), especially Figure 4 in the post, captioned: "Anthropogenic Climate Change." The figure is a 1486 AD woodcut of a "witch" conjuring up a hailstorm with the jawbone of an ass (donkey). Even in 1486, the public thought bad people made bad weather. Then they were witches, and now they are employees of fossil fuel companies. The logic and the data supporting the fantasy hasn't improved in 500 years.

The absurdity of humans controlling climate was even recognized in 800 AD when Archbishop Agobard of Lyons presented a sermon entitled "On Hail and Thunder," which contains the following:

"In these parts nearly everyone – nobles and common folk, town and country, young and old – believe that human beings can bring about hail and thunder…We have seen and heard how most people are gripped by such nonsense, indeed possessed by such stupidity…"

Again, not much has changed.

Jumping way ahead of our data, even when done with computer models, and proclaiming that climate change is man-made and dangerous before we see any direct evidence is dangerous. This will cost us in standard of living, and it will reduce the rate of improvement in poverty, malnutrition and income inequality that we have enjoyed for decades. But, most importantly, it will reduce our ability to adapt to climate changes.

Further, asserting climate is dangerous, using dodgy climate models to project climate into the far future, then taking the output from the models and running it through dodgy economic models to compute costs, is a strange and dangerous fiction. The one takeaway from this chapter that I hope every reader sees is that we really do not know what the climate future is, and we certainly cannot compute the cost. Further, no one credible is predicting any serious costs due to global warming, whether natural or man-made, for over 100 years. We have time to wait and see what is going to happen before we impoverish ourselves.

The IPCC only seems to focus on the negative aspects of climate change and not the positive aspects. The positive aspects are most apparent today, increasing crop yields, increasing arable land and fewer dying due to cold. The benefits of global warming outweigh the hazards today and will for the foreseeable future.

Chapter 3 Summary

We have seen that, if human carbon dioxide emissions are causing most of the current global warming, halting or reducing the use of fossil fuels is more expensive than adapting to the warming. This is because all adaptation measures require energy, and fossil fuels are currently the cheapest energy available for most purposes. Hydroelectric and nuclear fission can be cheaper for producing electricity, but electricity is only a portion of our energy needs and only 36 percent of the energy used to make electricity is delivered in a usable form for the customer.

Currently, popular carbonless energy sources, such as wind, biomass and solar use more energy than they produce and are not practical for most people. The current cost of reducing our fossil fuel use is very high and will likely get higher over the next 100 years. No one will see a benefit from this expenditure before the 24th century. This makes the present value of such an expenditure very small.

Projections of the cost of global warming and the warming itself are not precise. The potential cost of curtailing or eliminating fossil fuels cannot be calculated with any precision either, but it is clearly very high and for an uncertain outcome over 100 years from now. The purported dangers of global warming assume that the warming we've seen in the modern world, since about 1850, is unusual. In particular, the IPCC has focused on warming since 1950 when fossil fuel use accelerated worldwide. Just how unusual is the recent warming? In the next chapter we investigate this.

CHAPTER 4

Global Temperatures in the Holocene

Much has been made of the estimated 0.8°C to 1.2°C of surface air temperature warming since 1900. Plots of global average surface temperatures from NASA and the University of East Anglia Climate Research Unit working with the UK Met Office Hadley Centre (usually abbreviated as HADCRU) are shown in Figure 29. The plots end in 2017 immediately after a strong El Niño weather pattern that always causes a strong uptick in global temperatures. So, is this increase of about one degree in any way unusual?

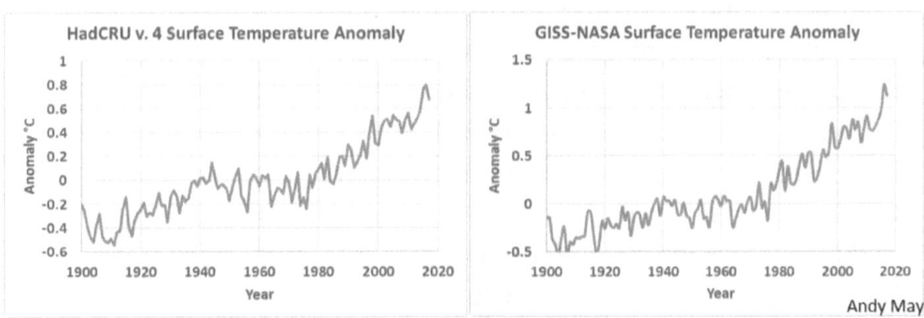

Figure 29. Global surface temperature records, (Met Office Hadley Centre and Climatic Research Unit 2017) and (NASA 2017b).

The first attempt to claim it is unusual was by Michael Mann when he published his now infamous "Hockey Stick" paper (Mann, Bradley and Hughes 1998) in *Nature*. This paper is often abbreviated MBH98. The hockey stick temperature reconstruction showed a long period of stable temperatures from 1400 to 1900 and later versions took it back to 1000AD, followed by a strong uptick at the end in the 20th century. This made the reconstruction look like a hockey stick. This hockey stick trend was an artificial construct created using deeply flawed statistical techniques, as described in (McIntyre and McKitrick 2005). For further criticism of the hockey stick the interested reader is referred to (May 2015c) or the *MIT Technology Review* article by (Muller 2004). The hockey stick and the data that went into it are so flawed as to be useless for representing recent temperature history. The following summary of the events following the publication of McIntyre and McKitrick's (M&M) critique of the hockey stick is by Ross McKitrick and from a 2018 legal complaint. See (McKitrick 2018) for the full text.

"After publishing their 2003 E&E [Energy and Environment] article and reviewing Mann's unpublished responses to it, McIntyre and McKitrick [M&M] submitted an extended critique of the errors and misrepresentations in MBH98 to *Nature* magazine, which had published the first of the hockey stick papers. *Nature* solicited a response from Mann et al., and after examining it they ordered Mann et al. to publish a detailed correction and restatement of their methodology, which appeared in June 2006. M&M also extended their critique of Mann's statistical methodology and submitted it to GRL [*Geophysical Research Letters*], which had published the 2nd hockey stick paper, and after peer review GRL published their study. Mann et al. never submitted a response. A panel led by Professor Wegman later conducted an independent review of the mathematical and statistical issues and upheld the M&M critique. A panel of the National Academy of Sciences also conducted an examination of the whole issue of paleoclimate reconstructions and upheld all the technical criticisms M&M made of Mann's work, going so far as to publish their own replication (North et al., 2006, pp. 90-91) of the spurious hockey stick effect M&M identified."

In this chapter we will examine another flawed Holocene temperature <u>record</u> by (S. A. Marcott, J. D. Shakun, et al. 2013) that we believe to be salvageable. In addition, we examine two widely used Greenland ice core air temperature records.

Holocene climatic overview

The Holocene is normally defined as "…the first signs of climatic warming at the end of the Younger Dryas/Greenland Stadial 1 cold phase…" (Walker, et al. 2009). It is often considered a geological epoch or series. It is part of the Quaternary geological period and equivalent to the older geological term "Recent." According to nearly all geological data, the Holocene began roughly 11,700 BP (in this book BP means before 1950AD). The first part of the Holocene is called the Holocene Climatic Optimum (HCO) or the Holocene Thermal Optimum (May 2015d). This worldwide climatic event started around 10,000 BP and ended before 4,000 BP. It shows up in nearly all worldwide and regional temperature reconstructions. The only significant exception is Antarctica. Around 4,000 BP the world started to cool, and the period from then until the present is called the Neoglacial. The coolest period in the Neoglacial, and in the entire Holocene, is the Little Ice Age (LIA) (May 2016f), which ended in the 1800s. Coincidentally, thermometer records of global temperature began just as the Little Ice Age was ending. Thus, alarmist references to "pre-industrial temperatures" are referring to a historically very cold period (Amos 2017).

The Marcott, et al. reconstruction (see Figure 34) shows a fairly flat Holocene Climatic Optimum temperature anomaly of +0.4°C from 9500 BP to 5000 BP, declining to a low of -0.4°C about 300 BP (1650 AD) in the Little Ice Age (LIA). This 0.8°C difference between the

HCO and the LIA is smaller than the generally accepted difference of 1°C to 1.5°C. The higher global temperature difference is clear in glacial records, as shown by (Koch, Clague and Osborn 2014). It can also be seen in the biosphere, as shown by (Kullman 2000), (Pisaric, Holt and Szeicz 2003), (M.MacDonald, et al. 2000), (Tinner, Ammann and Germann 1996) and (Thouret, et al. 1996). Further, the marine biosphere also shows a larger temperature difference, as seen in (Werne, et al. 2000) and (Rosenthal, Linsley and Oppo 2013). Rosenthal, et al. present very good evidence that the North Pacific water was 2.1° ±0.4°C warmer than in the 20th century and Antarctic water was 1.5° ± 0.4°C warmer. The Holocene Climatic Optimum and the Little Ice Age are the major climatic features in the Holocene, present in almost all reconstructions, and the sea surface temperature difference between them is well documented worldwide. Marcott's reconstruction can be rejected on this basis alone.

This is a new look at Marcott's proxies. Some recoil at any mention of Marcott, et al. because of the problems pointed out by (Foster 2013), (S. McIntyre 2013c) and others. But, we should not throw out the baby with the bathwater. Except for the problems noted above, it is a good collection of proxies and most of them are marine sea surface temperature proxies. This is a good thing because most of the thermal energy or heat capacity on the Earth's surface is in the oceans. As we recall from Chapter 1, over 99.9 percent of the surface heat content is in the oceans and only 0.071 percent is in the atmosphere. Warming of the atmosphere is not particularly significant on a climatic scale, so to study the last 12,000 years of climate change we should focus on the oceans as much as possible.

Ocean temperatures and their significance

Surface air temperatures have only been measured globally in a standard way since about 1880. Most of these temperature measurements have been made in the lower few meters of the atmosphere and on land. Ocean temperatures are typically measured just below the surface of the ocean and are called sea-surface temperatures, or SST. Argo floats (Argo 2017) and some buoys, like the Japanese TRITON (NOAA 2017) buoys, measure temperatures in deeper water. Oceans cover 70 percent of the Earth's surface and contain nearly all the surface thermal energy. For these reasons, when studying climate, the rate of change in the temperature of the oceans is more important than the rate of change in the temperature of the atmosphere.

We can see in Figure 30A (this is also Figure 1 in Chapter 1) that the temperature of the oceans has been increasing about 0.003°C per year since 2004 according to the JAMSTEC ocean temperature grid. If this rate of change continues, it amounts to an ocean surface temperature change of 0.3°C in a century. This increase might be real, but we cannot tell because the accuracy of the Argo floats that collect the data is no better than ±0.005°C and the accuracy of the JAMSTEC map grid is no better than ±0.2°C. The data and error are examined in (May 2016a).

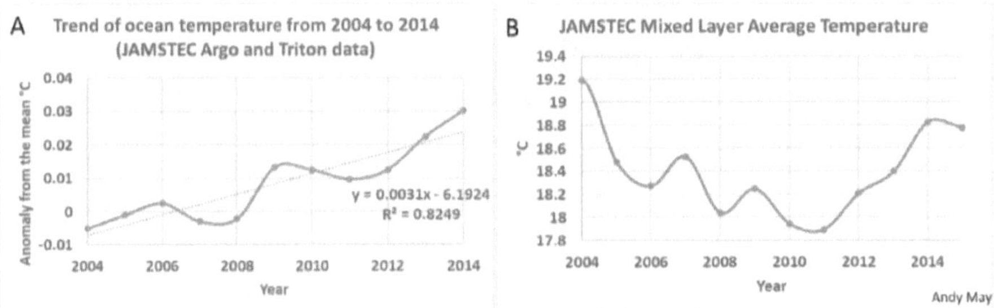

Figure 30. Average ocean temperature and the average ocean mixed layer temperature, data sources: JAMSTEC (May 2016a) (JAMSTEC 2016).

The upper 60 meters of the ocean interacts directly and quickly with the atmosphere. This layer is called the "mixed layer." Figure 30B plots its global average temperature. Due to the constant interaction with the atmosphere and turbulence caused by wind and wave action, the mixed layer maintains a fairly constant vertical temperature, density and salinity. Figure 30B shows no identifiable trend over the period where we have good data, except a slight decrease of a half degree Celsius or so over 11 years. This decrease is not significant. The time period is too short.

Using data collected from ships, the UK Met Office/Hadley Centre has created the "HADSST" dataset of sea surface temperatures. This data is shown from 1900 to 2016 in Figure 31, where it is compared to the more commonly seen HADCRUT land/SST merged dataset. It is notable that the sea surface temperatures (SST) cool slower than the overall temperatures from 1945 to 1975, and they warm slower than the overall temperatures from 1975 to 2016. Thus, even over short periods of time, we can see that the oceans act to dampen global temperature changes.

Figure 30 covers the period from 2004 to 2014. The shape of the HADSST record in Figure 31 is not dissimilar to the trend shown in Figure 30 for this eleven-year period.

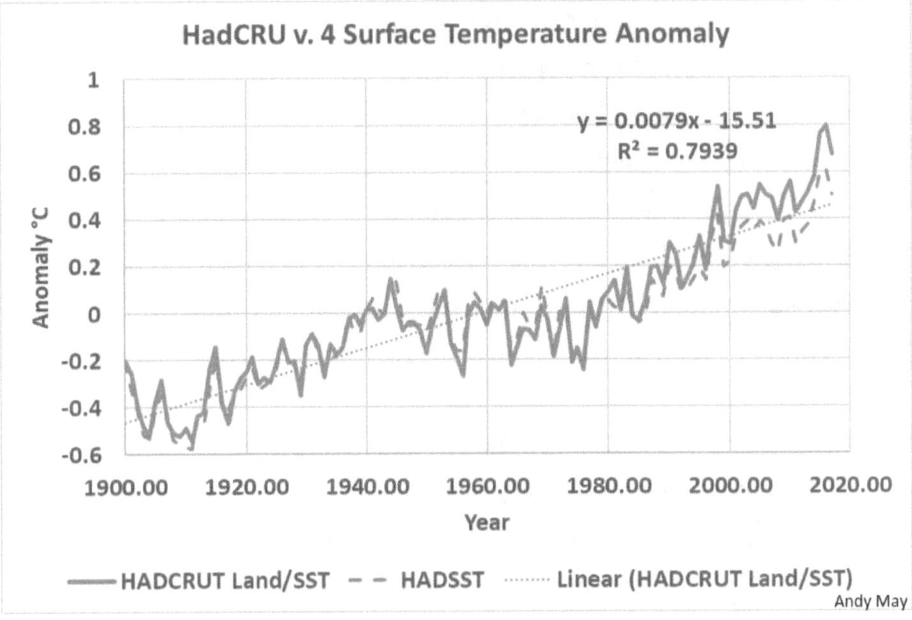

Figure 31. Global merged land/sea surface temperatures and global sea surface temperatures (SST), source (Met Office Hadley Centre 2017).

It is important to understand that the average global temperatures plotted in Figure 31 are not real, but they are useful. The temperature everywhere around the planet varies hour-by-hour. We try and take a few snapshots in a few places around the world and average them, but a true global average surface temperature is never achieved. If we are consistent from year to year (we aren't) then the difference from year to year in the temperature anomaly can tell us something, but with low accuracy.

Another consideration regarding ocean temperatures is critical to understanding how the oceans regulate global surface temperatures. Two large, very deep-water ocean masses, the Antarctic Bottom Water and the North Atlantic Deep Water, contain 55 percent of the total global ocean volume. This water is normally isolated from the surface, and it moves very slowly. It is ventilated and thermally equilibrated with the surface only in the polar regions. For this reason, the mean ocean temperature is tied to the polar regions, especially the Antarctic, which is surrounded by the Southern Ocean, the only deep continuous connection between all the oceans (Bereiter, et al. 2018).

The change in the global temperature anomaly does correlate with changes in the biosphere, especially with global changes in the mountain treeline. It also correlates with changes in mountain glaciers and sea ice. We scale temperature anomalies in degrees Celsius, but this is misleading since the connection to the physical Celsius scale is lost in the process of computing the global anomaly. More technically, the local temperature measurements are intrinsic measurements of temperature, but the conversion to a global anomaly creates an extrinsic value that is not rooted in the Celsius standard. Intrinsic properties are measured with an independent absolute scale and are independent of the surroundings. Extrinsic properties (like a global temperature anomaly) are only meaningful relative to other things or other anomalies measured in the same way.

Likewise, we scale paleo-temperature anomalies from proxies (see below for a discussion) in degrees Celsius, but this is also misleading.

The anomaly units for proxy and instrumental temperatures are dimensionless index values that try to approximate a true Celsius scale, and they are not necessarily comparable to one another. Here we will ignore this problem and try and compare them to get a qualitative idea of whether or not current warming is unusual. This is not ideal, but it is the best we can do with what we have. We will discuss some of the problems with the data and the error in the measurements below:

Much of the data presented in this chapter is from marine temperature proxies that reflect ocean temperatures from the mixed layer and just below the mixed layer. As we see in Figure 31, due to the large heat capacity of the ocean, these temperatures vary more slowly than air temperatures. The specific heat capacity of a substance, like water, is a measure of the amount of thermal energy it takes to raise the temperature of a kilogram of the substance one-degree Celsius. Water has a very high specific heat capacity, four times higher than the heat capacity of air. Further, the oceans are over 264 times as massive as the atmosphere, so it takes over 1000 times the thermal energy to warm the oceans one degree as it takes to warm the atmosphere one degree.

The atmosphere has 0.07 percent of the heat capacity of the Earth's surface, and the ocean mixed layer (the upper 60 meters or so) has 23 times that, or 1.6 percent. Marine sea surface temperatures (SST) are a good long-term indicator of climatic changes on the surface. This characteristic is also a downside since they miss, or "dampen," extremes that we might wish to see. Comparing them to true air temperature proxies, like the Greenland ice core data discussed in the next section, can give us an idea of how much they are dampened.

Greenland ice cores

For many years Richard Alley's <u>2004 GISP2</u> central Greenland Ice Sheet Project temperature reconstruction (Alley 2004), based on several central Greenland ice cores, was the most commonly cited Holocene temperature record (Alley 2000). There were good reasons why it was a popular temperature reconstruction: the data quality was excellent and the dates assigned to each annual ice layer were accurate. But, as shown <u>by</u> (Vinther, et al. 2009), the derived temperatures were affected by elevation differences that were due to episodic melting at the ice sheet margins, ice flow, and overall thinning over time since the last glacial maximum. Vinther, et al. provide us with the definitive Greenland Holocene temperature reconstruction shown in Figure 32 with the better known GISP2 (Alley 2000) record overlain as a dashed line for reference.

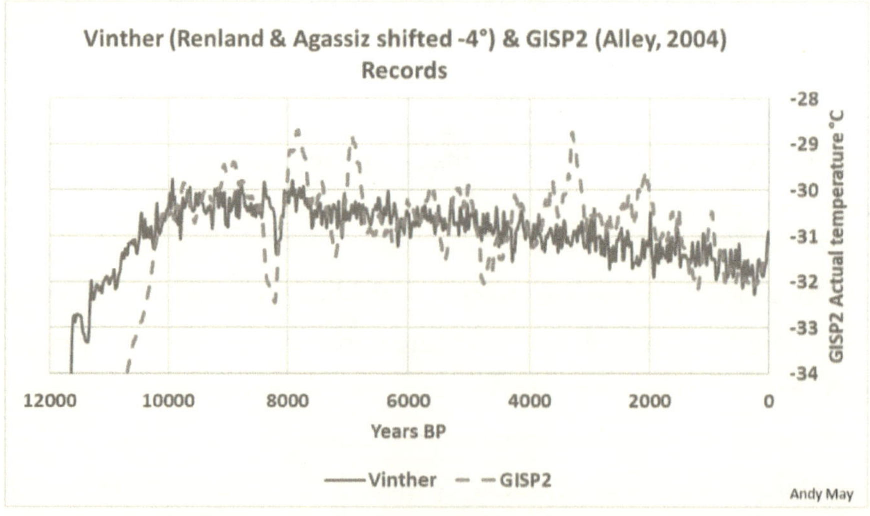

Figure 32. Vinther Greenland temperature reconstruction, elevation corrected, compared to the Alley GISP2 reconstruction, which is not elevation corrected.

The GISP2 core is from very high on the Greenland Ice Sheet and has a cooler average temperature than the coastal Renland and Agassiz ice cores (see Figure 33 for the core locations). So the Renland-Agassiz reconstruction by Vinther, et al., has been moved 4°C down to overlay the GISP2 record and make them easier to compare. The GISP2 and Vinther Renland-Agassiz air temperature reconstructions both show numerous examples of rapid (20 to 200 year) temperature changes of over one degree Celsius.

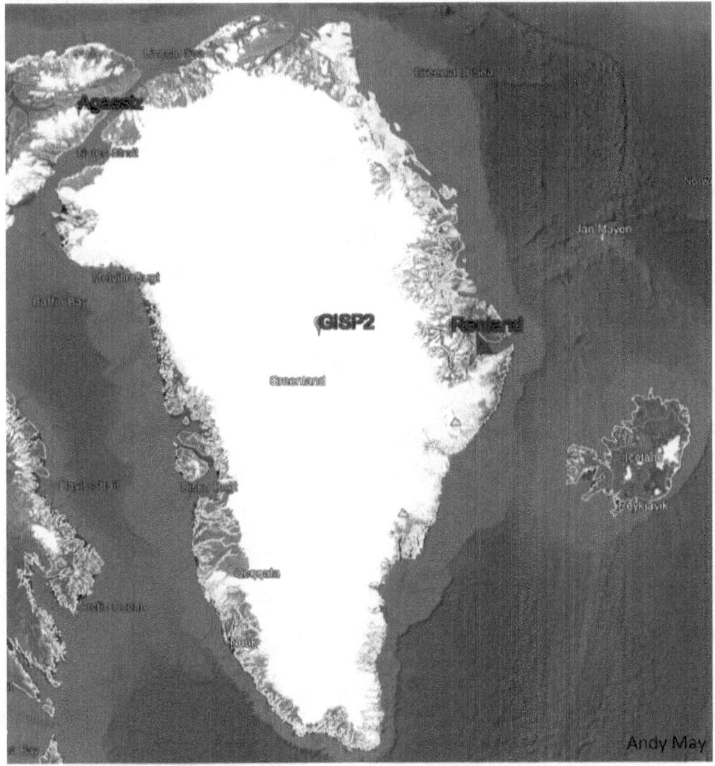

Figure 33. Google Earth Pro map of Greenland showing the locations of the ice cores discussed in the text, data from U.S. Department of State Geographer and Landsat/Copernicus Image IBCAO, used with permission.

There were two main problems with the GISP2 temperature reconstruction identified by Vinther, et al. (2009). The first was the lack of a prominent Holocene Climatic Optimum (HCO), which is seen in almost every other Holocene Northern Hemisphere and tropical temperature proxy. Further, the Agassiz and Renland temperature reconstructions, corrected for their well-documented elevation history, match each other but not GISP2, even though GISP2 sits in between them (see Figure 33). I refer the interested reader to <u>Vinther, et al., 2009</u> for the details. They also found that the total gas content data of the Renland and Agassiz ice core bubbles matched their elevation history. Generally, we believe that the Vinther, et al. reconstruction is the best Greenland temperature record and have used it in our Arctic reconstruction, rather than GISP2.

Greenland versus the rest of the world

The Vinther et al. reconstruction may be the best one for Greenland, but is it the best reconstruction to use for the entire planet? It is a reconstruction of surface air temperatures. This is a bit of a problem because the sea surface temperatures are probably more significant from a climatic perspective. Also, the North Atlantic weather might change in long term cycles that are out of sync with the rest of the world, especially the Southern Hemisphere. The only serious previous attempt to obtain a global temperature reconstruction was done by <u>Marcott, et al., 2013</u>. While Marcott and colleagues collected an impressive worldwide, mainly marine, set of temperature proxies, their analysis of them was deeply flawed and widely criticized.

While the changes in sea surface temperatures are more significant from a climatic perspective, we are attempting to compare a very high-resolution global surface air temperature/SST record, such as HADCRUT4 or GISS-NASA, to lower resolution paleo-temperatures from ice cores and marine proxies. The more proxies we include from around the world, the more global the reconstruction, but every proxy added, no matter how accurate, lowers the resolution. This is the inevitable result of averaging.

Marcott, et al. used some proxies with sample intervals as large as 500 years, and temperature variations in their reconstruction over periods shorter than 2,000 years are not statistically significant due to the statistical methods they used (S. A. Marcott, J. D. Shakun, et al. 2013). We rejected proxies with sample intervals larger than 130 years. The proxies with the best resolution have one sample every 20 years. This is still a coarse record, and comparisons to the very high-resolution and precise modern instrumental record must be made with this in mind.

Our high-resolution thermometer records since 1900 (see Figures 29 and 31), excluding strong El Niño years, show maximum changes of about 0.45°C over 20 years and a total change over the last 117 years of 0.8 to one-degree Celsius. The best fit land/sea surface temperature line shown in Figure 31 has a slope of 0.0079 degrees/year, suggesting an overall change of 0.9°C over 117 years. As we will see in this chapter, this high-resolution instrument global temperature record can be called "cherry-picked" since it begins immediately after the Little Ice Age .

How does this 117-year change compare with the maximum changes seen in the Greenland and Antarctic ice core records over the last 2,000 years? We've listed three recent extreme changes in Table 3.

Vinther, 2009 - Renland-Agassiz, Greenland				Pettit, 1999, Vostok, Antarctica		
Rapid temperature change examples				Rapid temperature change examples		
From	To	Change		From	To	Change
Year BP	Year BP	°C		Year BP	Year BP	°C
450	370	1.26		495	397	2.41
850	830	1.228		848	788	0.68
1980	1880	1.307		2009	1853	0.51

Table 3. Three recent periods of rapid warming in the Vinther reconstruction compared to comparable periods in the Antarctic Vostok reconstruction, data sources (Vinther, et al. 2009) and (Petit, et al. 1999). "Year BP" means years before 1950.

It's important to compare only ice core data to the modern land/sea surface record for two reasons. First, the ice core data is the highest resolution paleo-data we have, and second, it is generally considered the most accurate, in terms of dating and temperature (S. A. Marcott, J. D. Shakun, et al. 2013) (Alley 2000). The problem with ice core data is it is generally only found in the Arctic and Antarctic. In Table 3 we selected three large positive, recent temperature changes in the (Vinther, et al. 2009) Greenland temperature record and then located similar intervals in the Antarctic Vostok ice core record by (Petit, et al. 1999). The Greenland record has a 20-year spacing and the Antarctic record has a 20 to 60-year spacing, so they don't match up perfectly. Table 3 has only two geographic points, but we can see three times, in the last 2,000 years, where the temperature rise is as fast or faster in the polar regions than the rise we've witnessed globally in the last 117 years. If we go back to the beginning of the Holocene, the Central Greenland ice core reconstruction shows over 10°C of warming in less than 144 years from 11,755 BP to 11,611 BP. To see a plot of the incredible temperature-rise in 11,755 BP, see Figure 54 in Chapter 5. It occurs at the beginning of the Holocene, immediately after the Younger Dryas cold period.

Comparing modern instrumental temperature records to ancient temperature proxies is complex and difficult. The comparison is obviously qualitative, but necessary if we are to determine if modern warming is simply natural climate variability or truly "climate change." For a more complete discussion of this topic, read Renee Hannon's post "Modern Warming – Climate Variability or Climate Change" (Hannon 2018). She concludes that modern warming is clearly in the range of normal climate variability.

Examining the Marcott, et al. proxies

The Marcott, et al. global reconstruction (see Figure 34) used many proxies that did not cover the entire period being reconstructed, thus as proxies come and go "proxy dropout" causes anomalies, such as the very noticeable spike at the end of the reconstruction, which is due to one point at 1940. Although Marcott, et al. had a total of 73 temperature proxies, only 25 of them had values within 50 years of 1940 AD, and those values average 0.06°C, considerably different from what we see in Figure 34.

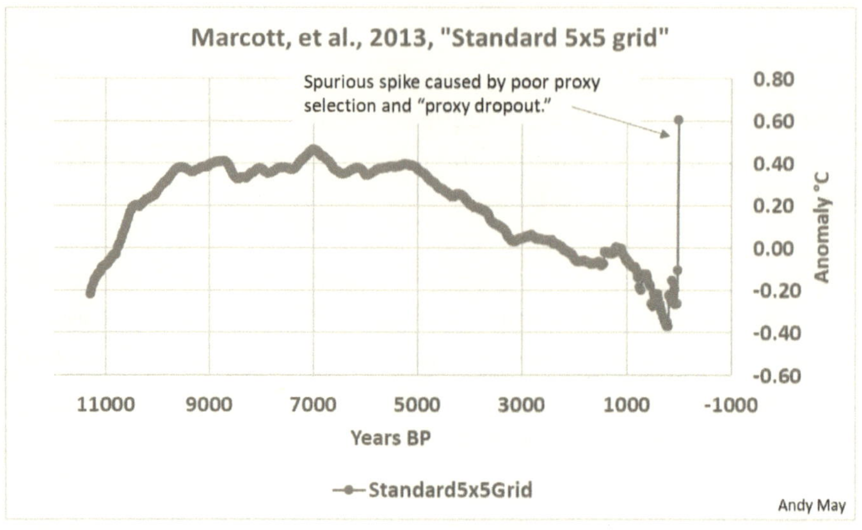

Figure 34. Marcott Holocene temperature reconstruction, data source Marcott, et al., 2013 supplementary material.

The spike is not real, is not in the underlying data, and should never have been included in the results or paper. The spike is due to proxy drop out, proxy inconsistencies and the authors changing some published dates in the proxies, according to Steve McIntyre at climateaudit.org (S. McIntyre 2013) (S. McIntyre 2013b) (S. McIntyre 2013c) and (Foster 2013) ("The Tick"). Even Marcott has acknowledged that his reconstruction from 1890 AD onward is not robust. The following quote is by Marcott and colleagues (S. A. Marcott, J. D. Shakun, et al. 2013b):

"We showed that no temperature variability is preserved in our reconstruction at cycles shorter than 300 years, 50 percent is preserved at 1000-year time scales, and nearly all is preserved at 2000-year periods and longer. Our Monte-Carlo analysis accounts for these sources of uncertainty to yield a robust (albeit smoothed) global record. Any small "upticks" or "downticks" in temperature that last less than several hundred years in our compilation of paleoclimate data are probably not robust, as stated in the paper."

That being the case, the tick at the end should never have been in the paper. Another problem is that 300 years is a very long time. Many very significant climatic events begin and end in less than 300 years. The 8.2 kiloyear event (May 2015b) is just one example. Marcott et al. (2013) chose too many proxies with poor temporal resolution, which, combined with an overly complex statistical procedure, reduced the resolution of their reconstruction to the point that significant historical details were lost. Table 4 lists the characteristics of the Marcott proxies and the proxies we selected for our reconstruction of Holocene temperatures.

Category	Number
Marcott's proxies	73
Added Rosenthal	1
Rejected proxies due to local conditions	4
Did not cover interval from LIA to HCO	5
Interval between samples > 130 years	34
Used in our reconstruction	29
4 proxies were combined into 2	2
Total	74

Table 4. Temperature proxy summary

Marcott used 73 proxies in his reconstruction. We added (Rosenthal, Linsley and Oppo 2013) and rejected four due to anomalous local conditions. The four rejects are explained more fully below. The two major climatic features of the Holocene are the Holocene Climatic Optimum (HCO) and the Little Ice Age (LIA). These are the warmest period in the Holocene and the coldest period, respectively. We rejected five proxies that did not cover enough of both extremes. Finally, we rejected any proxy that had a sample interval of more than 130 years. This left us with 29 proxy records after combining four partial proxies into two records.

Proxy selection

It is very logical to investigate long term climate changes using ocean temperature proxies from foraminifera shells and fossils ("forams"), planktonic and algal material. Traditionally the magnesium and calcite (Mg/Ca) percentages or $\delta^{18}O$ (the ratio of oxygen-18 and oxygen-16, see $\delta^{18}O$, Wikipedia) ratios in planktonic (floating) foraminifera have been used to deduce ancient sea surface temperatures. On land, ice core $\delta^{18}O$ records are used to estimate ancient air temperatures. Pollen is created by flowering plants and is very durable, so it is well preserved in lake sediments and peat. Pollen grains have distinctive shapes that can be used to identify the type of plant they came from. Certain plants are very temperature sensitive, so the relative abundance of different types of pollen can be used to infer ancient air temperatures (Brewer, Guiot and Barboni 2014).

More recently we have seen more use of haptophyte algal alkenones, especially the $U_{37}^{K'}$ (also written as UK'37) index to get ancient sea surface temperatures (Pahnke and Sachs 2006). TEX_{86} records from marine plankton have also been used recently to obtain sea surface temperatures and are among Marcott's proxies (Huguet, et al. 2006). The original reference for the TEX_{86} sea surface temperature proxy is (Schouten, et al. 2003). A complete list of the proxies used by Marcott, plus one we added from (Rosenthal, Linsley and Oppo 2013), and references and links to the original papers can be downloaded from (May 2017d).

The proxies were examined considering the criticism of Marcott's analysis by (S. McIntyre 2013c) and (Foster 2013). The proxies used in our reconstruction were selected using the following criteria:

1. The span of the proxy reconstruction had to cover at least 8,000 BP to 600 BP so the proxy covered part of the LIA and the HCO.

2. The resolution (time between samples) had to be less than 130 years.

3. Complex statistical techniques were avoided as much as possible.

The major climatic features of the Holocene are the Holocene Climatic Optimum (HCO) and the Little Ice Age (LIA), and, as noted above, there is abundant evidence that the global temperature difference between these two points exceeds 1.2° Celsius. Therefore, it seems logical to make sure the proxies cover at least part of both periods. Further, since all the reconstructions are temperature anomalies, we have built the anomalies as differences from the proxy mean between 9000 BP and 500 BP.

There is concern that the reason the Marcott, et al. (2013) reconstruction is underestimating the LIA to HCO temperature difference is because they included too many proxies with very long sample intervals. Proxies with long sample intervals miss essential detail and smooth and dampen any reconstruction. This can be made worse if Monte Carlo analysis techniques are used, since averaging many Monte Carlo "realizations" further smooths the result. We will use simple averages and fewer proxies to attempt to retain more of the variability, the peaks and valleys. The dates assigned to individual samples do not match, so all proxies were fitted with a cubic spline and resampled to the same 20-year sample increment. This was to preserve as much detail as possible.

Climate changes over the Holocene occur, in large part, by latitude. This is due to the Earth's orbital obliquity (the amount of axial tilt, which changes over a 41,000-year cycle) and precession (wobbling of the axis over 19,000 to 23,000 year cycles), as well as the long-term transport of thermal energy by ocean currents. As the Earth precesses or wobbles about its axis, it changes the hemisphere of the Earth that faces the Sun when the Earth is closest to the Sun (perihelion) or farthest from the Sun (aphelion). These changes affect the seasonality, the difference between summer and winter, in each hemisphere. Seasonality was maximal in the Northern Hemisphere at the beginning of the Holocene, as can be seen in the Northern Hemisphere summer insolation curve in Figure 35 (N-JJA). Much warmer summers helped kick off the melting of the continental glaciers.

The effect of obliquity is explained well by (Vinós 2017). Obliquity was also maximal at the beginning of the Holocene. Obliquity and precession worked together to bring us out of the last glacial maximum. Obliquity controls the amount of insolation received at the poles, relative to the amount received in the tropics. In the same blog post, Vinós explains this orbital effect in this way:

"Changes due to **obliquity have the effect of redistributing insolation between different latitudes** following an obliquity cycle of 41,000 years. When obliquity was maximal 9,500 years ago, both poles received more insolation due to obliquity, while the tropics received less. Obliquity also affects seasonality, at maximal axial tilt, there is an increased difference between summer and winter at high latitudes. But unlike precession changes, **obliquity alters the amount of annual insolation at different latitudes** in a 41,000-year cycle. This is represented by the background color of Figure 34 [our Figure 35], that shows how the polar regions received increasing insolation from 30,000 yr BP to 9,500 yr BP. Since then, and for the next 11,500 years, the poles will be receiving decreasing insolation. Unlike precessional insolation changes, **obliquity changes are symmetrical**. Although the annual insolation change is not too large, it accumulates over tens of thousands of years and the total change is staggering, creating a huge insolation deficit or surplus. This changes the equator-to-pole temperature gradient, and is largely responsible for entering and exiting glacial periods [(Tzedakis, Crucifix and Wolff 2017)] and for the general evolution of global temperatures and climate during the Holocene."

The emphasis is in the original blog post. Vinós' Figure 34 is presented below as our Figure 35.

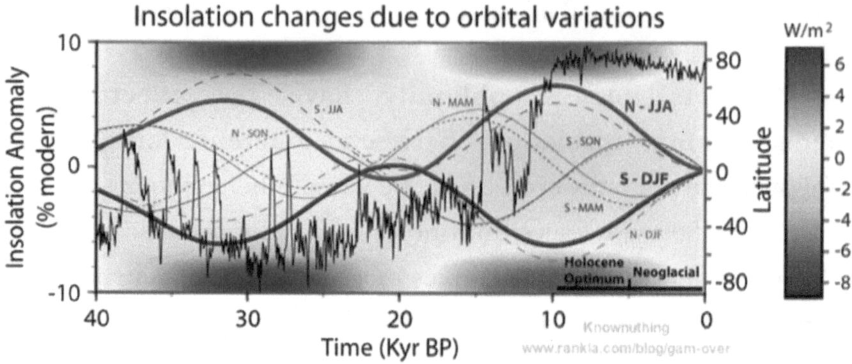

Figure 35. Insolation changes for the past 40,000 years. The background color represents the obliquity changes to the Earth's axis. The curves represent the changes to the total insolation at 60°N and 60°S. The beginning of the Holocene has high polar insolation and high summer insolation. Source: (Vinós, Nature Unbound III: Holocene climate variability (Part A) 2017), used with permission.

<u>Vinós' description</u> of Figure 35, in part:

"[our Figure 35]. **Insolation changes due to orbital variations of the Earth**. The insolation changes for the last 40,000 years are represented. Black temperature proxy curve represents $\delta^{18}O$ isotope changes from NGRIP Greenland ice core (without scale). The insolation curves are presented as the insolation anomaly for summer, winter, spring and fall. N (red) or S (blue) are the Northern or Southern Hemisphere, and the three letters are the month initials. Northern and southern summer insolation [are] represented with thick curves. Background color represents changes in annual insolation by latitude and time due to changes in the Earth's axial tilt (obliquity), shown in a colored scale. … The Holocene Climatic Optimum corresponds to [a] high insolation surplus in polar latitudes (red area), while Neoglacial conditions represent the first 5,000 years of a 10,000-year drop into a high glacial insolation deficit in polar latitudes (blue area)." (Vinós, Nature Unbound III: Holocene climate variability (Part A) 2017).

Since latitude has such a large influence on climate and climate changes, we will produce reconstructions in 30° regions of latitude. The Antarctic region, with latitudes from 90°S to 60°S, is presented first. Next, we present reconstructions for the Southern Hemisphere mid-latitudes (60°S to 30°S) and for the tropics (30°S to 30°N), then the Northern Hemisphere mid-latitudes and the Arctic. Our final global reconstruction is an area-weighted average of all regions.

In Figure 36, we compare Marcott's final reconstruction in orange-dashed to our final reconstruction, which uses fewer proxies and a simpler computational method. Our reconstruction is more active, has a similar HCO anomaly and a deeper LIA. The Marcott reconstruction has a difference of about 0.8°C between the HCO and the LIA, much less than documented in the literature. Our reconstruction has a difference of about 1.2°C between the HCO and LIA, in line with the above cited geological and biological evidence. Some well-known historical climatic events, such as the 8.2 kyr cold event, the 5.9 kyr event, the Roman Warm Period (2,000 BP), the 4.2 kyr event, and the Greek Dark Ages (also called the Greek Dark Age) are easily seen in our reconstruction but missing from the Marcott reconstruction. The abbreviation "kyr" is short for kiloyear and it means thousands of years before 1950. Both reconstructions capture the HCO and LIA, but the Marcott reconstruction dampens their amplitude.

The historical climatic events listed and marked in Figure 36 are discussed and documented in detail in (May 2015b). In summary, the 8.2 kyr event was a sudden cold event that ended the Pre-pottery Neolithic "B" period and caused a mass migration of people all over the world. The 5.9 kyr event was when the Sahara began to turn from a savanna to a desert. It is clearly seen in Figure 37.

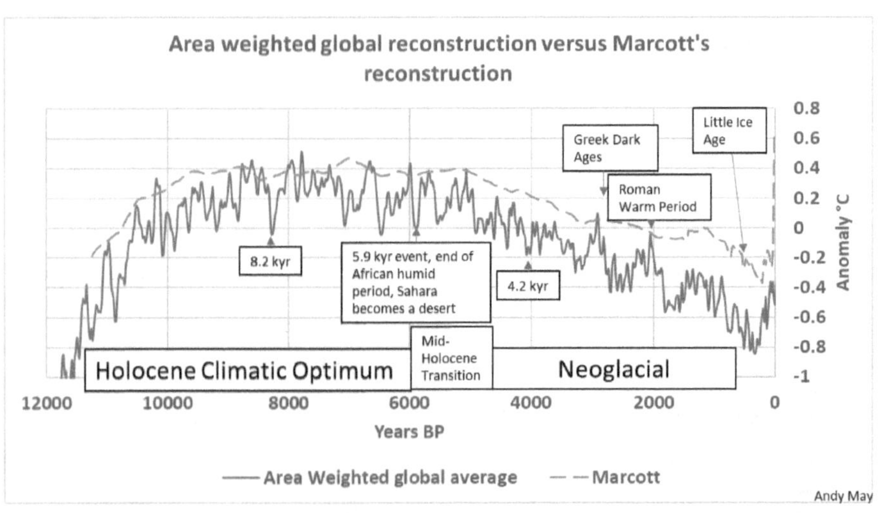

Figure 36. Marcott's final reconstruction (orange-dashed line) compared to our final reconstruction.

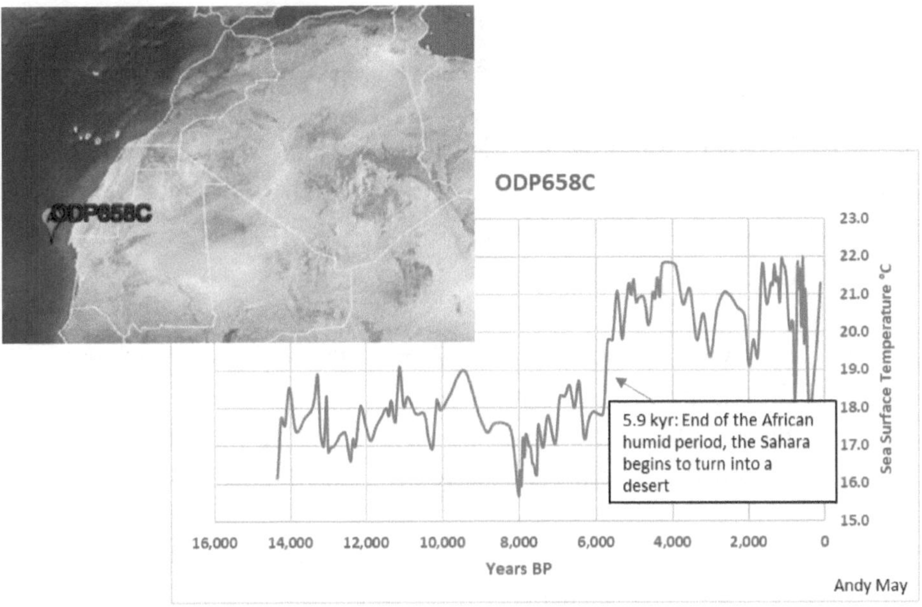

Figure 37. Plot of the temperature reconstruction of the ODP658C sediment core by deMenocal, et al., 2000 (Science). The core is from just offshore of the African Sahara Desert, as shown in the inset Google Earth Pro map. Map data from US Department of State Geographer, SIO, NOAA, U.S. Navy, NGA and GEBCO.

One of the five Marcott proxies we rejected due to local conditions was <u>ODP658C</u>, by (DeMenocal, et al. 2000). It is ideally located in the Atlantic, offshore of Western Sahara to record the 5.9 kyr event. It shows a sudden 3°C jump in sea surface temperature that begins 5,900 BP (see Figure 37) and completes 500 years later (±100 years). This abrupt change is thought to be due to the movement of the Intertropical Convergence Zone (ITCZ) from north of the Sahara to south of the Sahara and marks the beginning of the Mid-Holocene Transition (MHT) shown in Figure 36. This change was caused by a change in the Earth's orbital characteristics and insolation. Figure 35

105

shows that Northern Hemisphere insolation in the summer (labeled as N-JJA) is decreasing rapidly 5,900 years ago, but more importantly the Earth's orbital obliquity is decreasing, and this reduces the polar insolation and increases the equatorial insolation (background color in Figure 35). Obliquity was maximal (> ~24°) at the beginning of the Holocene Climatic Optimum but had reduced to < ~ 23.6° by the Little Ice Age. For a more complete description of the changes 5,900 years ago see (Vinós 2017), especially his Figure 39. The effects on civilization are discussed in (May 2015b).

The 4.2 kyr event was an abrupt cool event that coincides with the collapse of the Egyptian Old Kingdom (Lawler 2015), the Liangzhu culture in China, and the Akkadian Empire. It was a time of glacial advances and intense aridity in the Middle East, North Africa and India.

Roughly 3,100 BP (1150 BC) saw the end of the Bronze Age in the Mediterranean, and the Greek Dark Ages started shortly after (Cline 2015) and lasted about 400 years. This was followed by the Roman Warm period, which began around 500 BC and lasted into the new millennium. The Little Ice Age began around 1300 AD and did not end until after 1850 AD. The deepest and coldest part of the Little Ice Age occurred around 1650.

The Marcott, et al. (2013) reconstruction (the orange-dashed curve in Figure 36) is based upon a 5° by 5° global grid created using their proxies. Gridding data is an accepted way of spreading unevenly distributed data evenly over a map, but it can cause distortions. The distortions can occur due to isolated, extreme values or because the values gridded are incompatible. Sometimes gridding problems are simply due to data clustering, that is many values in one part of the map.

Table 5 shows the distribution of the proxies we have chosen by region. Marcott, et al. (2013) used 73 proxies for their reconstruction. We retained 28 of these and added the Rosenthal et al. (2013) Indonesian proxy for a total of 29.

Proxy Distribution				
Region	No. of Proxy Records	Area sq. km.	Area %	Proxy %
Antarctic	3	34,167,841	6.7%	10.3%
Southern Hemisphere (60S to 30S)	3	93,348,277	18.3%	10.3%
Tropics (30S to 30N)	7	255,032,236	50.0%	24.1%
Northern Hemisphere (30N to 60N)	7	93,348,277	18.3%	24.1%
Arctic	9	34,167,841	6.7%	31.0%
Earth	29	510,064,472		

Table 5. Distribution of the temperature proxies.

The proxies are much more numerous in the Northern Hemisphere than in the south, and for this reason, we chose not to grid the data, but instead create five simple latitude bounded regional reconstructions and merge them. The process is simple and easily reproducible. The five regional proxies are weighted using the surface area in each region to obtain a global reconstruction. This is not only simpler than gridding the data, but less likely to be affected by extreme values and spurious extrapolation. All calculations were done with R, a statistical software package that is available for free.

Summary of the reconstructions

Much more information, the R code, and data used can be obtained from the original blog posts. The original posts are: (May 2017e) <u>Part 1</u>, (May 2017f) <u>Part 2</u>, (May 2017g) <u>Part 3</u>, and (May 2017h) <u>Part 4</u>. Following is a summary of each reconstruction and the final global reconstruction.

Antarctic (90S to 60S)

Of necessity all the Antarctic proxies are ice core proxies and estimates of paleo air temperature. We chose the best three ice cores, and the anomalies from mean are shown in Figure 38.

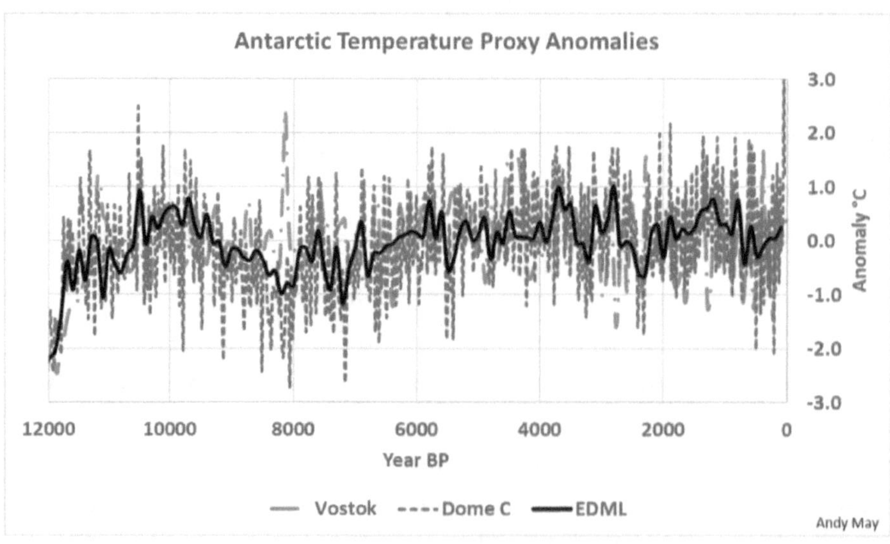

Figure 38. The Antarctic proxies.

The notable thing about these proxies is that the Antarctic temperature record is fairly flat for the Holocene, with the anomalies varying between -1° and 1°. The other odd thing about these proxies is that the temperatures drop between 10,000 BP and 6,000 BP, which is the opposite of what happens elsewhere.

The final Antarctic reconstruction is shown in Figure 39. The temperature drop during the HCO may be because winter (S-JJA) and fall (S-MAM) insolation are dropping in the Southern Hemisphere and summer insolation (S-DJF) is quite low at this time, as can be seen in Figure 35. Where the Neoglacial is a cooling trend in the tropics and in the Northern Hemisphere, it begins as a warming trend in Antarctica, followed by a flat trend. The distinctive historical events identified in Figure 39 still show up, but they are not prominent in this reconstruction, except for the 8.2 kyr event. Antarctica is somewhat isolated climatically, as it is surrounded by the Southern Ocean.

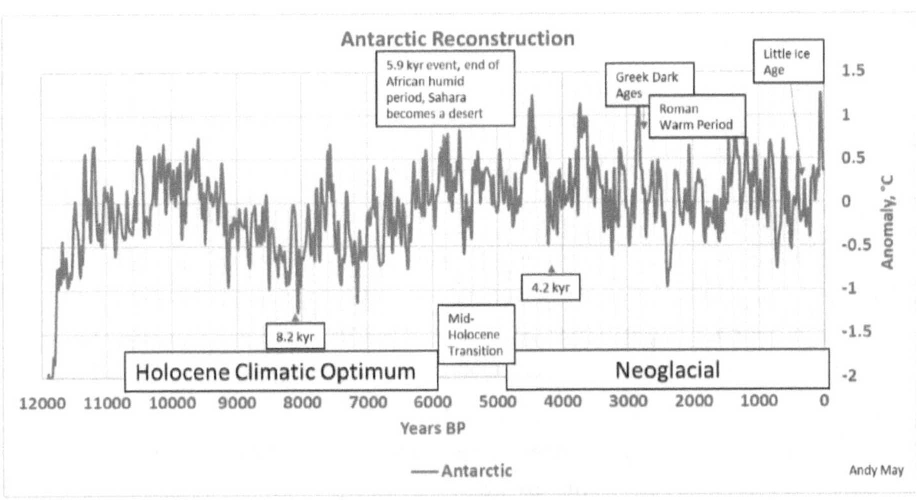

Figure 39. The Antarctic temperature reconstruction.

Mid-Latitude Southern Hemisphere

Unfortunately, we have little usable data between 60S and 30S, the mid-latitude Southern Hemisphere. Several proxies were rejected because they had large sample intervals, and <u>TN057-17</u> (Nielsen, Koc and Crosta 2004) was rejected because it is very anomalous. TN057-17 is a sea surface temperature proxy located in the Southern Ocean, right on the Antarctic Polar Front (APF). The APF is where a very abrupt sea surface temperature change occurs.

It is the southern limit of synchrony with the Northern Hemisphere climate system. The Antarctic Polar Front is a fluid boundary that changes seasonally but normally lies between 50°S and 62°S (Dong, Sprintall and Gille 2006). This is the only latitude band on the Earth where the ocean is uninterrupted by land all the way around the world. It is here that the colder Antarctic water dives below the warmer northern ocean water and forms a water mass called the Antarctic Intermediate Water (Talley 1996). The surface of the Southern Ocean in this region is dominated by the eastward flowing Antarctic Circumpolar Current that connects all the world's oceans and redistributes thermal energy (heat), salt and nutrients.

Currently, the TN057-17 location has sea ice cover about two weeks per year (<u>Nielsen, et al., 2004</u>). But, the duration of annual ice cover has changed a lot during the Holocene, and this probably had a dramatic effect on the proxy. The maximum sea ice cover was 4,300 BP, which is also the time of the lowest TN057-17 temperature. For more details see (May 2017f) or (Nielsen, Koc and Crosta 2004).

The three proxies that meet our criteria are plotted in Figure 40. They are from offshore Chile and New Zealand. The Homestead Scarp and Mount Honey records are primarily terrestrial pollen air temperature proxies. The GeoB_3313-1 and the MD97-2121 are two composited UK'37 marine alkenone proxies. They are from the same latitude, but <u>GeoB 3313-1</u> (Lamy, et al. 2002) is from offshore Chile and <u>MD97-2121</u> (Pahnke and Sachs 2006) is from New Zealand.

The proxies mostly agree after 5,000 BP, but before that they diverge. This is the regional reconstruction with the lowest confidence.

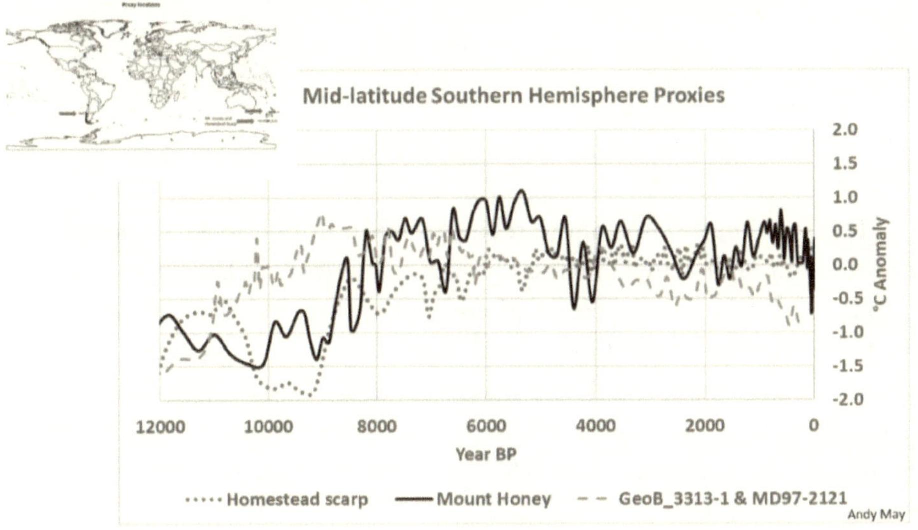

Figure 40. Mid-latitude Southern Hemisphere Proxies.

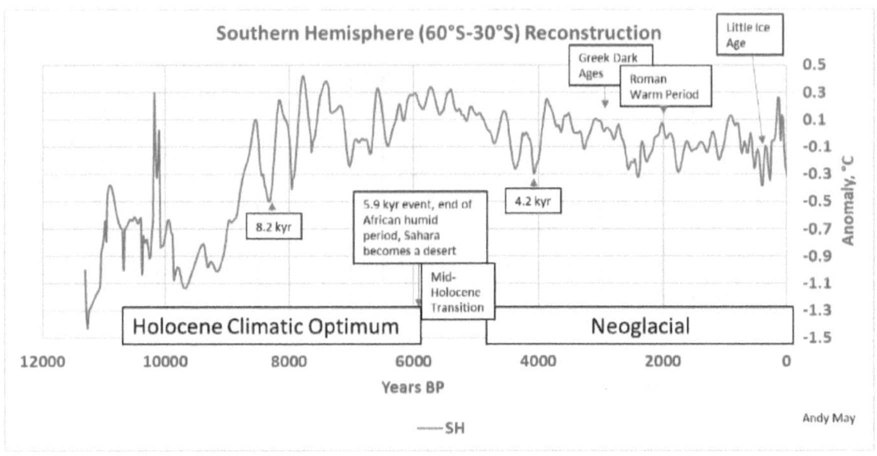

Figure 41. The Southern Hemisphere temperature reconstruction.

The Southern Hemisphere reconstruction in Figure 41 does show a prominent Holocene Climatic Optimum, but it starts later than in the tropics or the Northern Hemisphere. The historical markers, except for the beginning of the Greek Dark Ages and the 5.9 kyr event, occur about where they should. There is some mild cooling early in the Greek Dark Ages, but the severe cooling occurs later than it does in the Northern Hemisphere. The 5.9 kyr event is either missing or very subdued in this reconstruction.

Tropics

There are eight proxies that meet our basic criteria in the tropics. These are plotted in Figure 42. ODP-658C (DeMenocal, et al. 2000), off the coast of West Africa, was rejected and has already been discussed. It is radically affected by the movement of the ITCZ 5,900 BP. This is a local event that does not relate to average global temperature. Proxy 17940 (Pelejero and Grimalt 1999) is from the South China Sea and could also be considered slightly anomalous since in the Neoglacial period it trends warmer, rather than cooler. It also shows no Holocene Climatic Optimum.

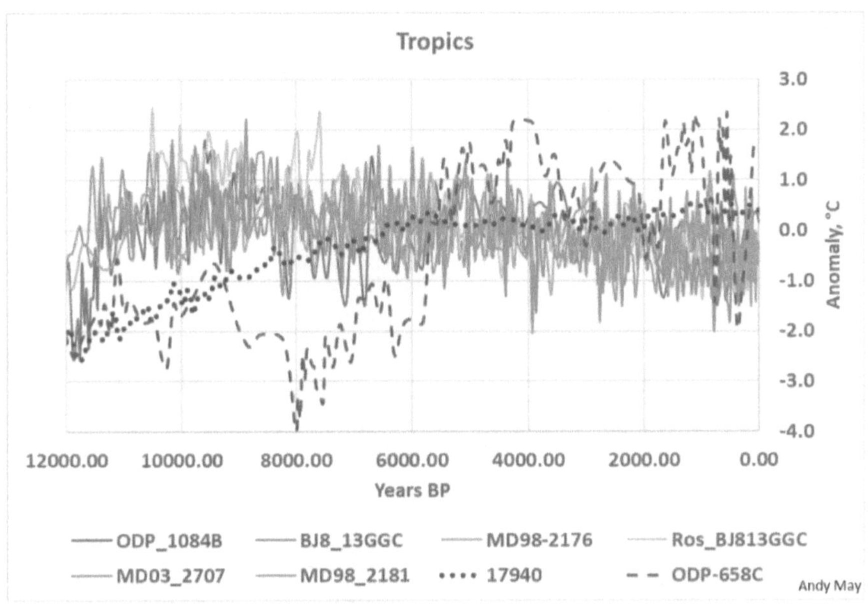

Figure 42. The tropical proxies.

The South China Sea is a very large marginal basin in the western Pacific. It is bounded by broad shallow shelves in the northwest and southwest that emerge during periods of low sea level. Sea level was low enough during the early Holocene for these shelves to be emergent. In addition, core 17940 is only 400 km from the mouth of the Pearl River, the second largest river in China. Due to the large sediment discharge from the river the location has a very high sedimentation rate. Further, it was larger in the past when the glaciers of the last glacial maximum were closer to the coast and sea level was lower. We kept the proxy in the tropical reconstruction but recognize that it is sensitive to the sea level changes experienced during the Holocene and to changes in the discharge rate from the Pearl River. The sea surface temperature proxies from other cores in the South China Sea are similar.

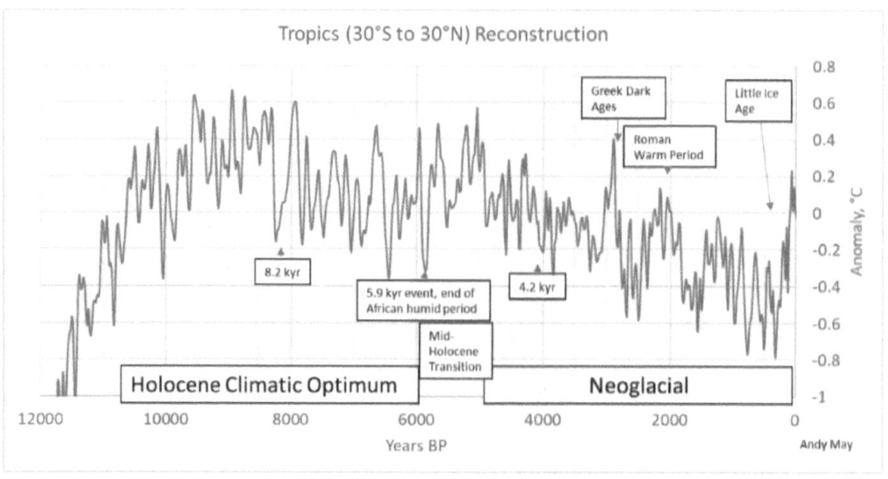

Figure 43. The tropical temperature reconstruction.

The tropics, as defined here, cover 30S to 30N and contain 50 percent of the surface area on the Earth. Thus, they represent the dominant region in our reconstruction. Figure 43 shows the final reconstruction. The HCO and the Neoglacial show up prominently, and all the major historical climatic events are clearly seen. The temperature scale is expanded on this plot because the variability in the tropics is low. This is because most of the surface is water and evaporation limits the maximum and minimum temperature.

Evaporation increases when temperatures rise (more on this in Chapter 7), and the resulting water vapor carries away latent thermal energy. The thermal energy is not released to the environment until the water vapor condenses as rain in a cooler location. Think of tropical storms or storm patterns that carry water vapor into temperate areas, such as Atlantic hurricanes or the west-to-east weather systems in the Northern Hemisphere.

The proxies that meet our basic criteria are in the Pacific, Indian and Atlantic Oceans, as shown in Figure 44. They are widespread enough that the reconstruction (see Figure 43) should be robust. All the proxies are marine proxies.

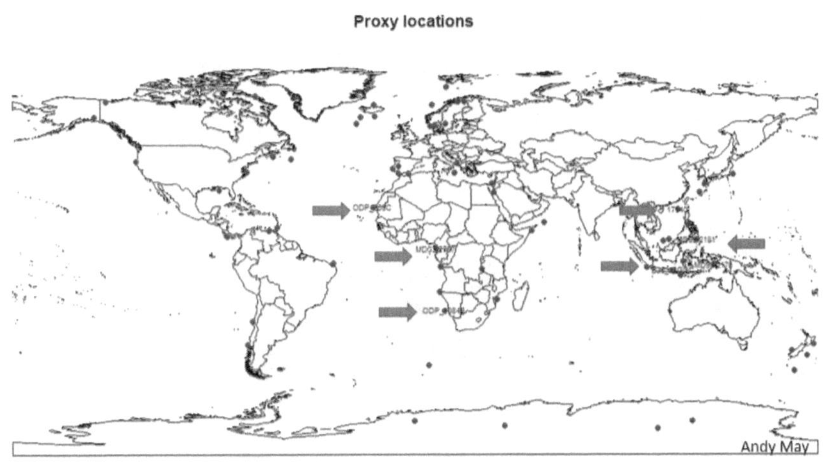

Figure 44. The locations of the tropical proxies are indicated by arrows.

Northern Hemisphere mid-latitudes

In all there are ten proxies that meet our basic criteria in the Northern Hemisphere, but two of them are merged into one. OCE326-GGC26 (Sachs 2007) and KY07-04-01 (Kubota, et al. 2010), just south of Japan, were anomalous and were removed from the reconstruction. The removed proxies were affected by river discharge, changing ocean currents, and changing sea ice cover. Northern Hemisphere rivers varied in their output as the continental glaciers melted and receded during the Holocene (May 2017g). The remaining seven proxies are shown in Figure 45.

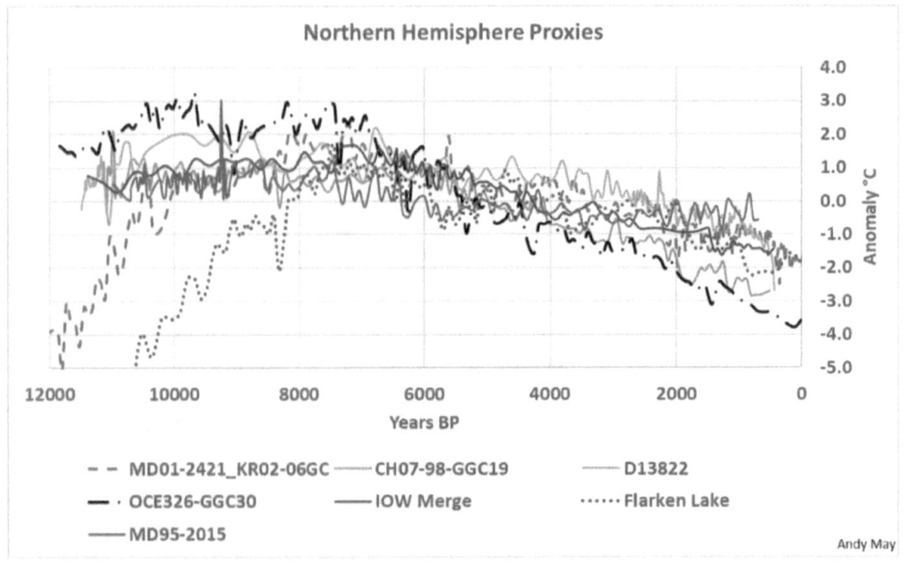

Figure 45. The Northern Hemisphere proxies.

All the proxies, except <u>Flarken Lake</u> (Seppa and Antonsson 2005), in Sweden, are marine sea surface temperature proxies. There is quite a spread of temperatures in the beginning since the warming after the Younger Dryas is delayed by the pattern of glacial melting in some areas. But after 8,000 BP the trends merge well. The locations of the Northern Hemisphere proxies are shown by arrows in Figure 46. There are proxies in both the Atlantic and the Pacific. The regional coverage is good.

Proxy locations

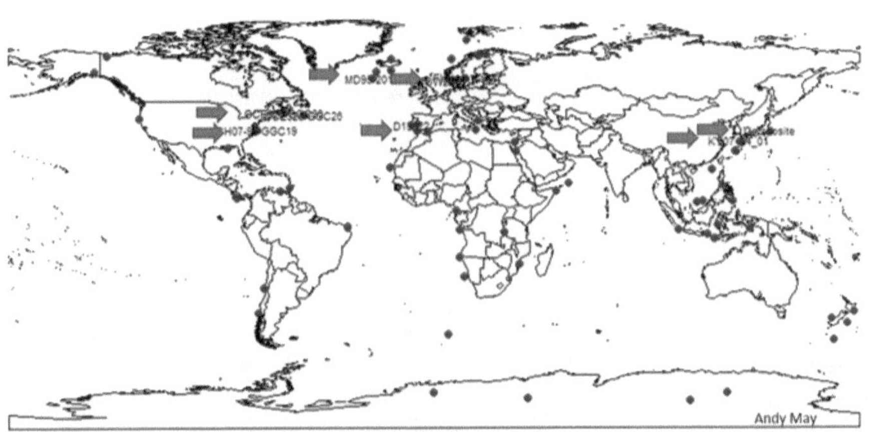

Figure 46. The locations of the Northern Hemisphere proxies

The final reconstruction is shown in Figure 47. The Holocene Climatic Optimum and the Little Ice Age are very prominent, and the difference between them is over 3.5°C. Even accounting for the scale change from the tropical reconstruction in Figure 43, some of the historical markers noted in Figure 47 are subdued. The 8.2 kyr, 5.9 kyr

117

events, and Roman Warm Period are about the same as in the Tropics, but the 4.2 kyr and the Greek Dark Ages have reduced amplitude.

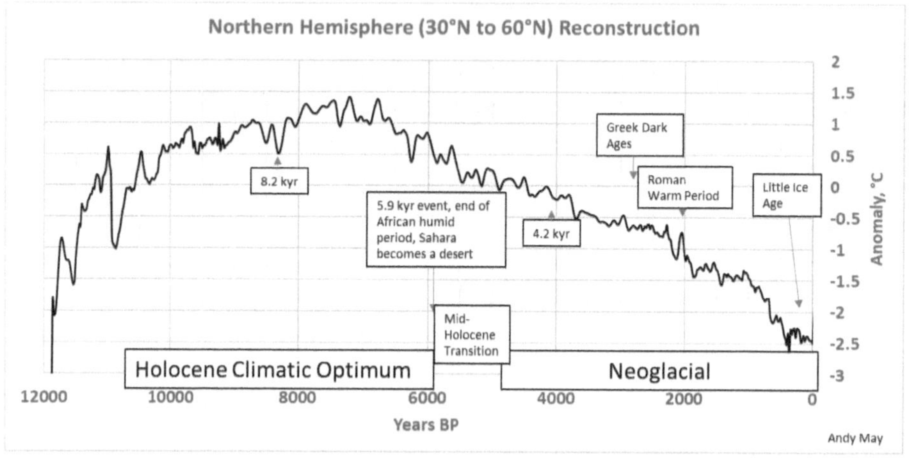

Figure 47. The Northern Hemisphere temperature reconstruction.

The Arctic

There are nine suitable proxies in the Arctic and all were used in the reconstruction. They are shown in Figure 48. Only <u>JR51GC-35</u> (Bendle and Rosell-Mele 2007) and <u>GIK23258-2</u> (Sarnthein, et al. 2003) look a little anomalous, but not severely so. GIK23258-2 is the most northerly proxy with a latitude of 75°N. This may explain the slightly anomalous warm anomalies at about 9,000 BP and between 2,500 BP and 1,000 BP. The Iceland proxy JR51GC-35 is quite spiky. JR51GC-35 is located north of Iceland in an area where multiple currents from the south and the north converge. This may explain the spiky nature of the curve. See the data source papers or (May 2017g) for more details.

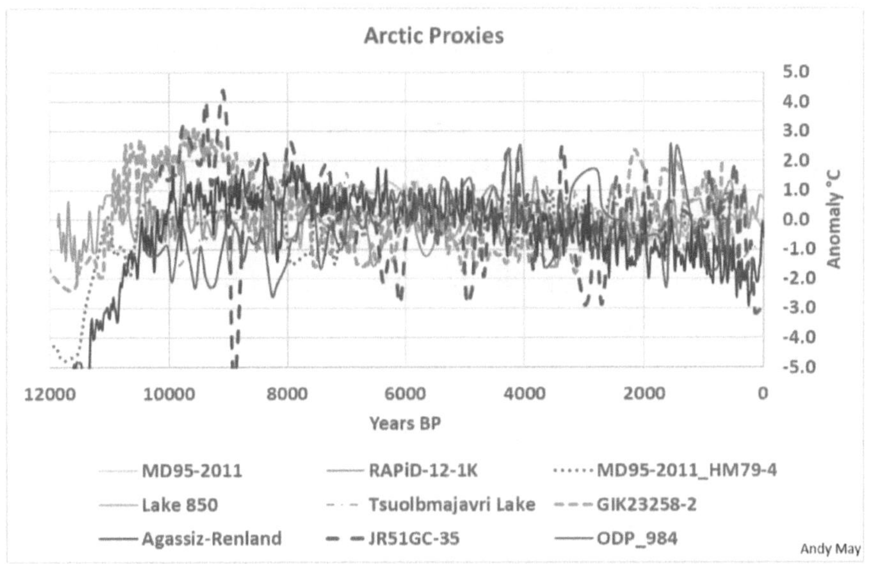

Figure 48. The Arctic proxies.

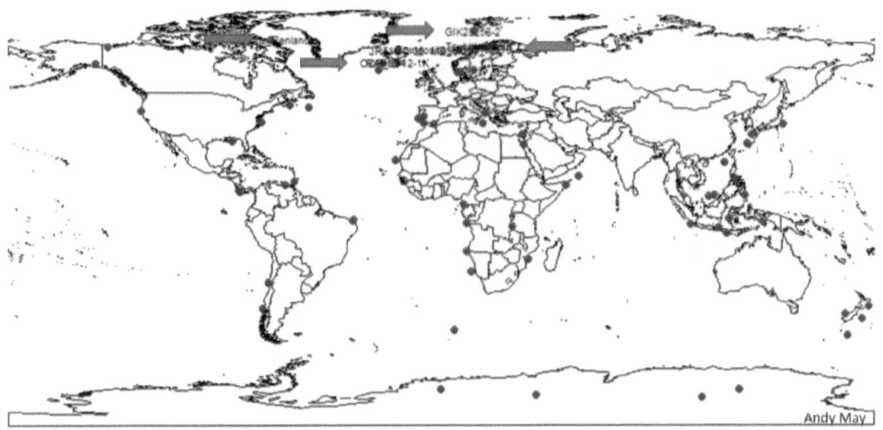

Figure 49. The Arctic proxy locations.

119

The proxies are all from the North Atlantic or the Arctic Ocean. Most of them are marine proxies. See Figure 49 for the locations.

The final Arctic reconstruction is shown in Figure 50. There is a noticeable, if somewhat subdued, Holocene Climatic Optimum and a very severe Little Ice Age. The 4.2 kyr and the 8.2 kyr events are quite prominent. The 5.9 kyr tropical event cannot be seen, but there is a slight cooling after this time, as the Neoglacial period is entered. The Greek Dark Ages can be seen.

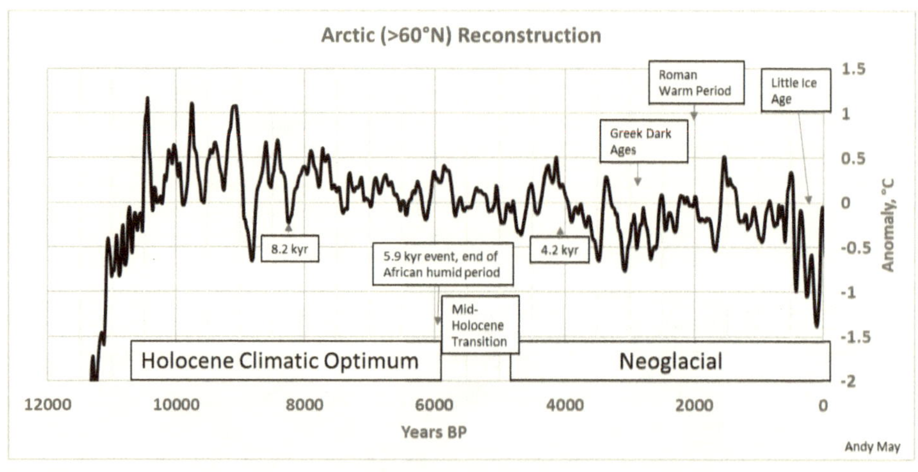

Figure 50. The Arctic temperature reconstruction.

The global reconstruction

In Figure 51 we see the five regional reconstructions. While all the regions, except for Antarctica, show a prominent HCO, it occurs at different times in different regions. The nature of the Little Ice Age also varies by region, but it occurs at roughly the same time everywhere.

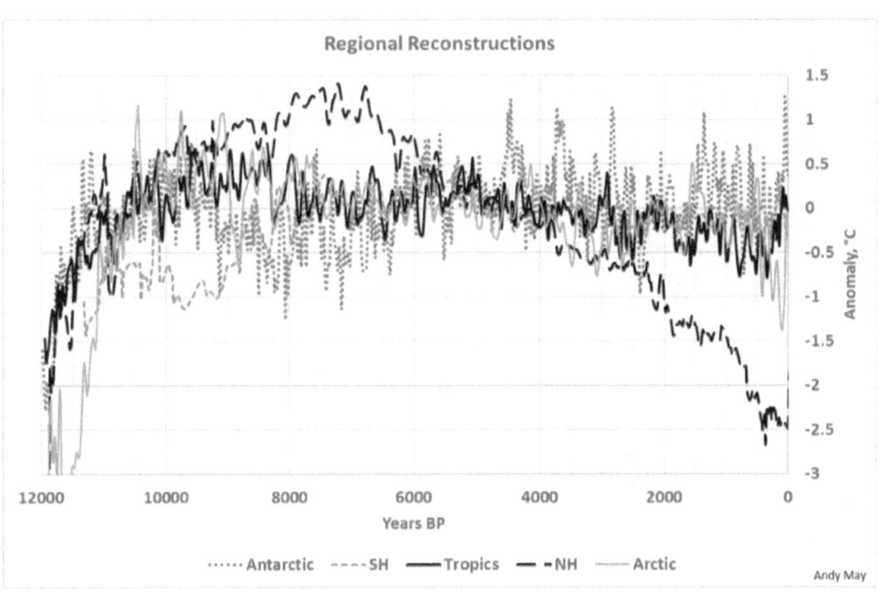

Figure 51. All the regional temperature reconstructions plotted together. The Northern Hemisphere reconstruction is the outlier.

It is interesting that the Northern Hemisphere is the odd reconstruction. This was also true for the Marcott et al. (2013) Northern Hemisphere reconstruction from 30°N to 60°N (see Figure S10f in their <u>supplementary materials)</u> (S. Marcott, et al. 2013c). The

Northern Hemisphere has the greatest temperature variation of the five regions and a clearly different trend. Is this because it contains most of the land? Perhaps so. It may be, in part, due to the impact of the melting continental glaciers from the last glacial advance. Certainly, the high Northern Hemisphere summer insolation early in the Holocene, due to orbital precession and obliquity, played a significant role (see Figure 35). At the beginning of the Holocene, the Northern Hemisphere summer had maximal insolation due to precession, and the higher latitudes (poles) had maximal insolation, due to obliquity, at the expense of the tropics. Thus, both the precession cycle and the obliquity cycle were in their warmest phases for the Northern Hemisphere mid- and high-latitudes. This changed a few thousand years later, and the climatic equator (the Intertropical Convergence Zone) shifted and the long Neoglacial cooling period began.

The Southern Hemisphere is also a bit anomalous, with a dip in the period of the HCO, corresponding with a dip in winter and fall insolation in the Southern Hemisphere, plus low summer insolation (see Figure 35). The other interesting thing about the reconstructions is that the Northern Hemisphere has a higher and longer Holocene Climatic Optimum. The Northern Hemisphere was affected much more by the last glacial advance due to the large continental ice masses there. The Southern Hemisphere ice was mostly sea ice, which, presumably, melts at a steadier rate with less dramatic effect.

Bereiter, et al. published a study of the mean global ocean temperatures for the past 22,000 years in *Nature* in January 2018 (Bereiter, et al. 2018). In it they try and show that the mean global ocean temperature is closely tied with Antarctic temperature. Our proxies are mostly marine proxies and do not contradict this hypothesis. Except for the beginning of the Holocene Climatic Optimum (9,000 to 7,000 BP), the Antarctic reconstruction follows the tropics, the Southern

Hemisphere and the Arctic fairly well. The Northern Hemisphere is the big anomaly.

The Arctic and Antarctic each cover 6.7 percent of the globe, the southern and northern mid-latitudes cover 18.3 percent each and the tropics covers 50 percent. If we weight each reconstruction by the area of its region we get the global reconstruction in Figure 52.

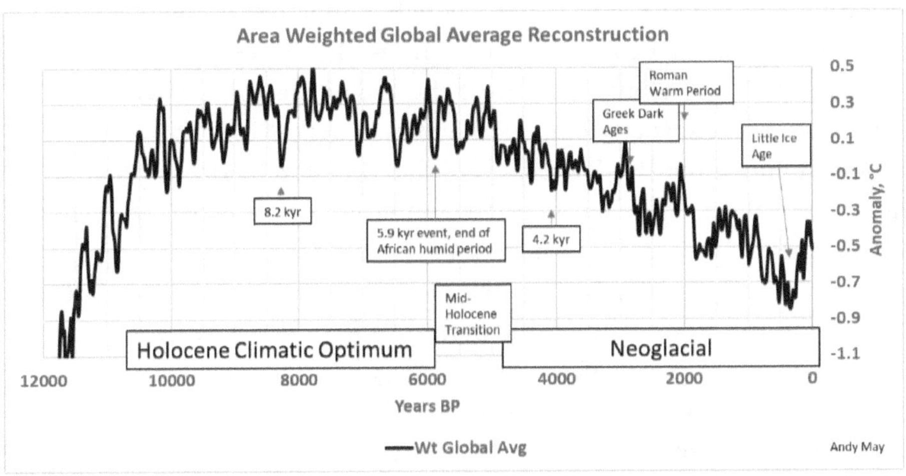

Figure 52. The final area-weighted global reconstruction.

In this reconstruction, the lowest temperature anomaly in the Little Ice Age (LIA), -0.85°C, occurs in 1610AD. The apparent Medieval Warm Period (MWP) is smeared over several hundred years and occurs from around 530AD to 1130AD, which does not fit the historical record. Oddly, only the Southern Hemisphere and the tropics show a distinct Medieval Warm Period (MWP) in its historical time. This is despite abundant historical evidence of a Northern Hemisphere MWP from around 900AD to 1200AD. The Antarctic reconstruction shows several warm spikes during the period, but nothing very distinct. The reason for the lack of a distinct MWP signature in the northern reconstructions is not known.

The Roman Warm Period (RWP) shows up well in the reconstruction at about the right time. The "collapse of civilization" (Cline 2015) at the end of the Bronze Age is clearly seen. The 4.2 kyr event that led to the collapse of the Akkadian empire in 4170 BP and the Egyptian Old Kingdom can be seen (deMenocal 2001). The 5.9 kyr event that occurred as the Sahara was turning into a desert, causing a great migration to the Nile valley, which ultimately resulted in the formation of the Egyptian Old Kingdom, is clearly seen. The LIA is the most significant climatic event of the Holocene without question, but the second most severe climatic event may well be the 8.2 kyr event. This event ended the Pre-Pottery Neolithic B culture and was possibly when the Black Sea was catastrophically connected to the Mediterranean in an event that may be remembered as Noah's great flood (Ryan and Pittman 2000). The 10.3 kyr event takes place about the time the Pre-Pottery Neolithic period began. For more details on human history and climate change see "Climate and Human Civilization over the last 18,000 years" (May 2015b). The historical climatic events match this reconstruction well, except for the MWP.

In this reconstruction, the depth of the Little Ice Age (LIA) occurs between 1430 AD and 1690 AD with an average temperature anomaly of -0.75°C. The peak of the Holocene Climatic Optimum (HCO) runs from 8760 BP to 7760 BP (6810 BC to 5810 BC). Ignoring the 8.2 kyr event, which falls in the middle of the warm period, the average anomaly is +0.4°C, with a peak as high as 0.51°C. The temperature difference between the LIA and the HCO is 1.15°C. This compares well to the geological and biological evidence summarized above and presented in (Vinós 2017).

A word about error

There are many sources of potential error in these reconstructions. We have emphasized the temperature proxies we thought were most important and significant. Specifically, we focused on the geographical distribution of the proxies, proxy selection, the choice of the mean used to generate the temperature anomalies, the effects of proxy dropout, proxy resolution, and the impact of local conditions on the proxies. The latter problem relates to how applicable the proxy is to regional climate as opposed to local climate. Examples of inappropriate proxies due to local conditions are TN057-17, ODP-658C, OCE326-GGC26 and KY07-04-01, as discussed above.

Dating the proxy samples can be problematic. (S. A. Marcott, J. D. Shakun, et al. 2013) emphasize potential dating errors in their paper and supplementary materials. They consider dating errors to be the largest source of error. Marcott, et al. (2013) also provide a very detailed discussion of proxy-to-temperature calibration uncertainty in their supplementary materials. Generally, they assume one standard deviation (normally distributed) to be the error inherent in the proxy-to-temperature conversion, otherwise they follow the proxy author's recommendations.

Marcott, et al. assumed a fundamental dating error of 120 to 150 years for most cases and accounted for it using a Monte Carlo procedure (1,000 realizations), which is explained in their supplementary materials. For the layer counted Antarctic ice-core records they assumed a ±2% uncertainty and for Greenland cores they assumed a ±1% error. Marcott, et al. recalibrated all radiocarbon dates using IntCal09 (Reimer, et al. 2009). Our reconstructions use the original published dates and not the recalibrated dates.

Dating errors and proxy-to-temperature errors are undoubtedly important. Marcott et al. (2013) provide a good discussion of these problems, and their supplementary database contains estimates for these sources of uncertainty. They also found that some of the proxies have a seasonal bias and attempted to account for this source of error in their Monte Carlo procedure. They do not believe that seasonal bias is an important source of error.

We have nothing to add to their work on these uncertainties and the interested reader is referred to their paper. They do present an interesting figure in their supplementary materials displaying the 1,000 Monte Carlo realizations that result from their study of error due to dating and proxy-to-temperature conversion. It suggests that error due to these factors is roughly ±0.5°C.

Marcott, et al. (2013) also provide their own latitudinal temperature reconstructions and display them in their supplementary Figure S10. Their regional reconstructions are different in detail than ours because they use more proxies, but their 30°N to 60°N reconstruction for the Holocene is the same big outlier we see in our Figure 51. They also note, as others have, that computer simulations of Holocene climate do not agree with the proxy reconstructions. This is often called the <u>Holocene temperature conundrum</u> (Liu, et al. 2014). The largest difference between the simulation results and the proxy reconstructions occurs in the mid-high latitude Northern Hemisphere, which suggests that the climate models are missing some key component of Northern Hemisphere climate. They suggest that the models may not be modeling North Atlantic Ocean circulation properly. We agree.

We believe the largest source of error in these reconstructions is in the proxy selection. As documented in this chapter, some of the original 73 proxies are affected by resolution issues that hide significant climatic events and some are affected by local conditions that have no global significance. Others cover short time spans and do not cover the two most important climatic features of the Holocene, the Little Ice Age and the Holocene Climatic Optimum.

Marcott, et al. attempted to account for error by determining a fundamental error in proxy sample dates and in the temperature estimates. They used a Monte Carlo procedure to vary the measurements in both date and amplitude to produce an "optimal" temperature reconstruction.

In our examination, we found that this procedure was misleading for two reasons. First, the potential errors in date and in temperature were large in comparison to the resulting reconstruction. This caused excessive smoothing of the result and a serious loss of important detail. As noted above, the error bars on their result are larger than the total span of temperatures in the reconstruction. Second, proxy selection was a larger source of error and this was ignored.

While some of the sources of error might be quantifiable, it does no good to account for the quantifiable errors with a Monte Carlo procedure, while ignoring the unquantifiable sources, such as proxy selection. This is especially true if the Monte Carlo procedure eliminates critical detail. Qualitative data, such as the historical markers shown on our reconstructions and the glacial and biological data we used to establish the ocean surface temperature difference between the HCO and the LIA, are critical and should not be ignored.

Chapter 4 Conclusions

We've tried to address the criticism of the (S. A. Marcott, J. D. Shakun, et al. 2013) global temperature reconstruction. Steve McIntyre,

Grant Foster and others criticized their adjustments of the published proxy dates, their inclusion of inconsistent proxies, and not compensating for proxy drop out. Their proxy reconstruction does not reflect abundant geological and biological evidence that the average sea surface temperatures were more than one degree Celsius warmer during the Holocene Climatic Optimum than during the Little Ice Age (Vinós 2017). In addition, the use of proxies that do not cover the interval from the LIA to the HCO is problematic since these are the two best defined temperature extremes in the period. Further, we are using temperature anomalies from the mean to build these reconstructions and prefer to get the mean from the period 9000 BP to 500BP so that the mean represents both the high temperatures of HCO and low temperatures of the LIA.

We rejected proxies with sample intervals greater than 130 years because they tend to reduce the resolution of the reconstruction and they dampen ("average out") important details. We did not use Monte Carlo simulation, like Marcott, et al. did, for the same reason. The final reconstruction shows a difference of 1.15°C between the LIA and the HCO. This suggests that the underlying data support this temperature difference.

By avoiding smoothing the data, we have captured more detail. Much of the additional detail appears to correspond to known climatic events. While the LIA, HCO, Roman Warm Period, the Greek Dark Ages, and other historical events show up well in the reconstructions, the Medieval Warm Period does not. It appears dampened and offset in time from historical records. The reasons for this are unclear. Perhaps the MWP occurred in different times or in different intensity in different places, smearing it on a global reconstruction? Either way proxy choice determines the MWP intensity and timing, which is disappointing. More work and better proxies are needed to improve our Holocene temperature record.

A precise Holocene temperature reconstruction is not possible. Even measuring the potential error in a reconstruction this long is incredibly difficult. Marcott, et al. (2013) did a good job of estimating dating error and proxy-to-temperature error, in our opinion. But, they do not address other important factors, such as historical events, glacial and biological data, proxy selection and resolution, which are probably more important.

Our final reconstruction is composed mostly of marine temperature proxies. The reconstruction shows many examples of very rapid changes of over 0.5°C. This is larger than the recent trend seen in the short partial instrumental record we have of the upper 2,000 meters of the overall ocean (Figure 30A). The portion of the ocean that interacts most with the atmosphere is called the "mixed layer." Over the short period that we have data for, the mixed layer shows no trend (see Figure 30B).

Air temperature changes tend to be faster and larger than ocean temperature changes due to the larger heat capacity of the ocean. This is easily seen in Figure 31. The short term (100- to 200-year) changes seen in the Vinther Greenland temperature reconstruction (Figure 32) are often larger than the 0.9°C change observed over the last 117 years. The HADCRU SST (sea surface temperatures) shown in Figure 31 have a trend of 0.7°C per century, as we come out of the Little Ice Age, the coldest period in the Holocene. This is comparable to the larger temperature swings in our proxy reconstruction shown in Figure 52. Taken all together, the climate change and/or global warming observed since 1900, or the late 19th century, until today does not seem unusual.

Chapter 4 Summary

The global, mostly shallow marine, Holocene temperature reconstruction (Figure 52) developed in this chapter shows many examples of temperature variability comparable to those seen in the instrumental period since the Little Ice Age. We recognize that the instrumental record and the proxy-based record developed in this chapter are not directly comparable and that this comparison is approximate and qualitative. The global proxy "degrees C index" is not the same as the global instrumental "degrees C index" and neither represent true intrinsic temperatures. But, we believe that this is the best we can do with what we have and accurate enough to demonstrate that the current warming is not outside the Holocene norm. To call modern global warming a problem and unprecedented requires showing it to be unusual. The data presented here shows, at the very least, it cannot be proven to be unusual.

Previous attempts to show it as unusual by (Mann, Bradley and Hughes 1998) and (S. A. Marcott, J. D. Shakun, et al. 2013) were deeply flawed technically and invalid. The instrumental record of global temperatures is short, only about 160 years, but the global average surface temperature has increased about $0.9°C \pm 0.1°C$ over that period (Met Office Hadley Centre and Climatic Research Unit 2017). The error from 1850 to 1950 is larger, roughly $\pm0.2°C$. From 1950 to the present, the error is about $\pm0.1°C$. The most accurate paleotemperatures we have are the Greenland and Antarctic ice core records. These records show several temperature rises larger than the current one in the past 2,000 years (Table 3).

Marine (sea surface temperature) variability is less than air temperature variability, but our low-resolution proxies show that global temperature variations of more than $0.5°C$ in 100 years are common during the Holocene.

The Southern Ocean is a deep-water ocean that connects all the oceans of the world. This ocean contains most of the Antarctic Bottom Water mass, which when combined with the North Atlantic Deep-Water, contain 55 percent of the total ocean volume. These deep-water masses are only ventilated to the surface in the polar regions. Thus, global ocean temperatures correlate well with Antarctic temperatures.

The regional temperature reconstructions show that the world did not warm uniformly during the Holocene, and the Northern Hemisphere warmed very differently than the rest of the world. Except for the early Holocene (9,000 BP to 7,000 BP) all the regions, except for the Northern Hemisphere, track the Antarctic reconstruction. This suggests two things: first that atmospheric carbon dioxide is not driving Holocene temperature patterns. Something else is at work, CO_2 is a global well-mixed gas and not isolated from the Northern Hemisphere. Second, the polar regions, especially Antarctica, have an outsized influence on ocean temperatures, which in turn affects atmospheric temperatures.

In this chapter we tried to show that the recent global warming is not unusual and that the pattern of warming is different from what would be expected from a global increase in CO_2. Global warming is not uniform around the world (see Figure 21 in Chapter 2), and it is concentrated in the Northern Hemisphere. The Southern Ocean and parts of Antarctica may not be warming at all.

As mentioned above and in Chapter 1, 99.9 percent of the Earth's surface heat capacity is in the oceans and less than 0.1 percent is in the atmosphere. Further, CO_2 is only 0.04 percent of the atmosphere. It beggars belief that a trace gas (CO_2), in an atmosphere that itself contains only a trace amount of the total thermal energy on the surface of the Earth, can control the climate of the Earth. This is not the tail wagging the dog, this is a flea on the tail of the dog wagging the dog. Extraordinary evidence is needed to convince us of this hypothesis. Since the impact of man-made CO_2 on climate has never been measured and is only crudely estimated with unvalidated models, the jury is still out on this idea.

Global warming and/or the rise of human civilization have sometimes been blamed for an increase in species extinctions. In the next chapter we examine these claims.

CHAPTER 5

Extinctions and the Gulf Stream

In this chapter we examine the assertion that man-made climate change, the growth of the human population, and other human activities are causing an increase in species extinctions. We also examine the polar bear controversy.

Finally, we examine the assertion that man-made climate change will cause an influx of fresh water into the North Atlantic from melting glaciers on Greenland and shut down the North Atlantic thermohaline circulation and/or the Gulf Stream. This will then, supposedly, cause a major cooling akin to the one seen 8,180 to 8,340 years ago when the ice dam holding Lake Agassiz-Ojibway onto North America broke, spilling a huge amount of fresh water into the North Atlantic.

Global warming is causing a "great extinction" event

Besides the potential financial costs of global warming, some claim that the modest 0.9°C of warming we have experienced since the mid-19th century is causing more species to become extinct. Others say it is the growth of the human population. In fact, *National Geographic* has claimed we are in the "sixth great extinction in the Earth's history" (Drake 2015). Nadia Drake cites a recent study (Ceballos, et al. 2015) that proclaims that the current extinction rate may be 100 times the normal rate. This is also the subject of a book entitled *The Sixth Extinction*, by Elizabeth Kolbert (Kolbert 2014). If true, this is a considerable cost of either global warming or human population growth. Some try to draw a semantic distinction between "mass"

133

extinctions and "great" extinctions, but generally the two terms are used interchangeably, as in the National Geographic article, and reference the Earth's five mass extinction events (BBC 2014).

Comparing the modern rate of species extinction to the pre-human civilization rate is complicated by many factors. First, there is no universally accepted definition of "species." Some classifications rely on morphology and some on the capability of the members to interbreed. Of necessity, paleontologists use the morphological definition since they rely on fossils. Paleontologists rarely work at the level of species, preferring to work on higher taxa, such as genus, family, order or class. If a genus goes extinct, the paleontologist will estimate the number of species that have gone extinct (Mora, et al. 2011). Some modern biologists use the interbreeding criteria (Mayr 1974). Mayr writes the following:

> "Darwin's choice of title for his great evolutionary classic, *On the Origin of Species*, was no accident. The origin of new 'varieties' within species had been taken for granted since the time of the Greeks. ... 'Descent with modification,' true biological evolution, could be proved only by demonstrating that one species could originate from another." (Mayr 1974) page 10.

> "A species is a protected gene pool. It is a Mendelian population that has its own devices (called isolating mechanisms) to protect it from harmful gene flow from other gene pools." (Mayr 1974) page 13.

> "The study of long-term evolutionary phenomena is the domain of the paleontologist. He investigates rates and trends of evolution in time and is interested in the origin of new classes, phyla and other higher taxa. Evolution means change

and yet it is only the paleontologist among all biologists who can properly study the time dimension. If the fossil record were not available, many evolutionary problems could not be solved; indeed, many of them would not even be apparent." (Mayr 1974) page 9.

Thus, we see the complexity of comparing the modern rate of species extinction with the overall geological "norm." Currently, there may be as many as 8.7 million species of plants and animals (Mora, et al. 2011). The generally accepted geologically recent rate of species extinction, the "background rate," is about 0.1 to 1 species extinction per million species per year (E/MSY). A few researchers think it might be as high as 2 E/MSY (Ceballos, et al. 2015).

Media reports often confuse the issue by treating sub-species or populations as if they were species. A species of a modern, living animal is defined as a "protected gene pool," as Mayr has defined it above. If it goes extinct, the gene pool is extirpated and lost. Sub-species and populations can interbreed and be recreated. Much has been made in the press about the Eastern cougar becoming extinct (Zuckerman 2018). But, the Eastern cougar is simply a population of the sub-species North American cougar or Puma concolor cougar. This is not an extinction but a restriction in the range of one sub-species, the North American cougar.

It is common today, and unfortunate, that the term extinction is considered equivalent to restricting the range of a population. Strictly speaking, the term "extinction" should be reserved for species and higher taxa (Gittleman 2017). When a population or sub-species disappears the species gene pool is still viable.

There is uncertainty in the number of species and in the number of extinctions. Thus, it is common to simply assume or estimate that the background extinction rate is one to two species per year. Given the wide uncertainty in the two numbers used in the calculation, this is a reasonable simplification and we will use it here.

This rate is based upon observed extinctions of genus or higher order taxa in the geological record and an estimate of the average number of species in each genus. Currently, we rely on observed or estimated extinctions. The observations are often based upon historical records of species that no longer exist, like the dodo bird. Indirect estimates have also been made, based on rates of deforestation and on species-area relationships (Ceballos, et al. 2015). Of necessity, only vertebrate extinctions are counted on land, since these form fossils. Ocean extinctions are monitored with both vertebrate fossils and animals with exoskeletons or shells.

The area for larger wild animals is decreasing as humans use more land for agriculture. This limits the area that can be used by wildlife, especially larger predators that can endanger people or domestic animals. However, man does set aside parks and other nature preserves for wildlife and often protects endangered species, and this is a relatively recent thing. So, is man causing a mass extinction?

The direct estimates of recent extinctions are limited to the period from 1500AD to the present and rely on written records. They usually are based upon records of a few hundred mammal or other vertebrate species extinctions, which are then compared to the total number of taxa in that group. This assumes that this ratio can be compared to the total number of species in the world, estimated to be up to 8.7 million. Or, at least, comparable to the taxa (mostly marine) counted in the past five great (or mass) extinctions.

We also must consider that the period from 1500 to 1850 covers the worst of the Little Ice Age, the coldest period in the Holocene and a time of great environmental stress. One study that claims we are in a human-caused "sixth mass extinction" states that:

"The evidence is incontrovertible that the recent extinction rates are unprecedented in human history and highly unusual in Earth's history." (Ceballos, et al. 2015).

One of the co-authors of Ceballos, et al. is Paul Ehrlich, who famously predicted England would cease to exist by the year 2000. The claim we are in a mass extinction is clearly based on a very small data sample gathered over a very short (geologically) time period and of dubious quality.

Experts in the past five mass extinction events do not support the idea that we are in a sixth mass extinction event. Smithsonian paleontologist Doug Erwin, an expert on the Permian extinction that may have wiped out 90 percent of the species on Earth, says those claiming we are in a similar situation simply do not know what a mass extinction is (Brannen 2017). Erwin points out that somewhere between 0 and 1 percent of species have gone extinct in recent human history. Consider what he told *The Atlantic*, in an interview:

"So, you can ask, 'Okay, well, how many geographically widespread, abundant, durably skeletonized marine taxa have gone extinct thus far?' And the answer is pretty close to zero," Erwin pointed out. In fact, of the best-assessed groups of modern animals—like stony corals, amphibians, birds and mammals—somewhere between 0 and 1 percent of species have gone extinct in recent human history. By comparison, the hellscape of End-Permian mass extinction claimed upwards of 90 percent of all species on earth." (Brannen 2017)

Daniel Botkin (UC Santa Barbara environmental scientist) has estimated that the extinction rate for animals and plants is about one per year (Botkin 2016). This is not an alarming rate and is very far from a "mass extinction" event.

"In the Danish press I pointed out that we had long been hearing figures for the extinction of the world's species which were far too high – that we would lose about half of all species within a generation. The correct figure is closer to 0.7 percent in 50 years. This led to the Danish chairman of Greenpeace, Niels Bredsdorff, pointing out that Greenpeace had long accepted the figure of 0.7 percent." Lomborg, Bjørn. *The Skeptical Environmentalist: Measuring the Real State of the World* (p. 17).

"Extinctions are always happening. Most species that have existed on Earth have gone extinct. As I mentioned before, the average rate for animals and plants has been about one a year (although some recent scientific papers argue that the average rate has been much higher, maybe up to six per year). The number of extinctions has varied over time, including five "mega-extinction" events since the evolution of multicellular life-forms, about 550 million years ago. During each of these mega-extinctions, the majority of species appear to have gone extinct in a (geologically speaking) comparatively short time. The greatest mass extinction was the Permian-Triassic, about 250 million years ago, when an estimated 80 to 90 percent of all species went extinct." Botkin, Daniel B. *25 Myths That Are Destroying the Environment: What Many Environmentalists Believe and Why They Are Wrong* (Kindle Locations 645-650).

The IPCC WGII AR5 report supports Lomborg and Botkin, in a bit of a back-handed way, with Figure 53 below from page 43 of the Technical Summary. It is a portion of their Figure TS.2 (Field, et al. 2014):

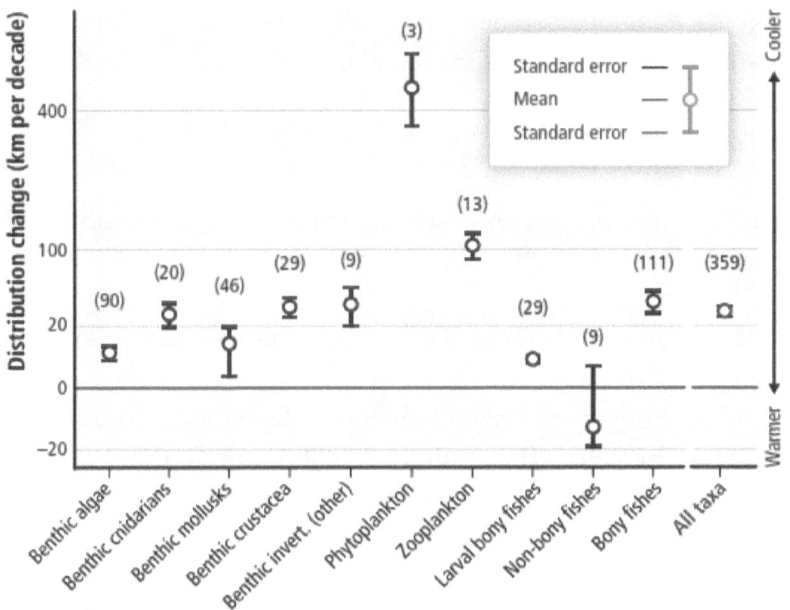

Figure 53. Rates of change in distribution (km/decade) for several marine taxonomic groups for 1900 to 2000. Source IPCC WGII AR5 Technical Summary, page 43, Figure TS.2, used with permission. Note: Positive changes mean the taxa are present over a larger area, generally moving poleward.

While the IPCC AR5 report does not claim a "great extinction" or that "mass extinction" is imminent, they do claim that a "large" fraction of both terrestrial and freshwater species face an increased risk of extinction due to climate change. This prediction is entirely model-based, and nothing like they predict has been observed. The report also acknowledges that there is very low agreement concerning the fraction of species at risk (AR5 WGII, technical summary, page 67).

The "Great" Quaternary megafauna extinction

Another claim of a recent "great" extinction is discussed by (Barnosky 2008). He claims that two-thirds of mammal genera and one-half of species that weighed more than 44 kg went extinct in the Quaternary megafauna extinction. This extinction occurred between 50,000 and 3,000 years ago.

After cleverly reducing his sample size to get to two-thirds, he proclaims:

"The Quaternary Megafauna Extinction (QME) killed >178 species of the world's largest mammals, those weighing at least 44 kg (roughly the size of sheep to elephants). More than 101 genera perished. Beginning ~50,000 years (kyr) B.P. and largely completed by 7 kyr B.P., it was Earth's latest great extinction event."

So, the "Great" Quaternary megafauna extinction event extirpated about 178 species of the world's largest mammals, those weighing >44kg. As noted above, according to (Mora, et al. 2011), there are 1.2 million catalogued species, and there may be as many as 8.7 million species alive in the world today. So, while the extinction of 178 species over ~50,000 years of rapid climate change is distressing because the creatures are so interesting, geologically it is not a significant extinction event. It is not remotely comparable to the five great mass extinctions, although Barnosky's language attempts to suggest they are comparable.

A "Mass Extinction" is defined by the (American Museum of Natural History 2017) as an event where more than half of all species go extinct in a short time. The museum's web site defines the most recent mass extinction in this way:

> "Around 65 million years ago, … Fossils that are abundant in earlier rock layers are simply not present in later rock layers. A wide range of animals and plants suddenly died out, from tiny marine organisms to large dinosaurs."

> "Species go extinct all the time. Scientists estimate that at least 99.9 percent of all species of plants and animals that ever lived are now extinct. So, the demise of dinosaurs like *T. rex* and *Triceratops* some 65 million years ago wouldn't be especially noteworthy--except for the fact that around 50 percent of all plants and animals alive at the same time also died out in what scientists call a mass extinction."

The Quaternary megafauna extinction event began ~50,000 years ago and was mostly complete by 3,000 years ago. This was an extinction rate of 0.004 species per year, which is much lower than the average long-term extinction rate of one to two species per year (Botkin 2016) (Ceballos, et al. 2015). Obviously, the extinction of sabre toothed tigers, dire wolves, mammoths and other megafauna is not particularly significant in geological history. But, these are fascinating creatures due to their size, and we value them much more than a toad or insect that goes extinct. However, this is a value judgement, not a geological event like a "mass extinction" or a "great extinction."

Barnosky notes that humans fit into the large mammal category, yet humans have survived and thrived. He also claims that the gain in human biomass largely matches the loss of large non-human megafauna biomass until 15,000 years ago, then total megafauna biomass crashed as many megafauna went extinct. As can be seen in Figure 54, this coincides with the abrupt Younger Dryas cooling event. Humans survived this brutally cold period, in part, by eating many of the megafauna that are now extinct. Our angst over these extinctions may be classic "survivor guilt." 12,000 years ago was about when human civilization began. Rice was already being cultivated in China and grains in the Levant (present-day Syria and Israel). The construction of the large stone monuments at Gobekli Tepe in modern-day Turkey began about this time. Humans were beginning to settle down and become farmers. Humans adapted quite well both during and after the Younger Dryas, but other megafauna did not.

Land is coveted by humans, and with the growth of civilization the concept of private property, especially for farming and ranching, was developed. Humans tend to protect their property and often fence it off. Thus, it is possible that the current problem for megafauna is not "biomass," as suggested by Barnosky, but space. Wild megafauna need a lot of space to survive. Humans prefer domesticated megafauna, like horses and cattle, to wild megafauna like wolves and coyotes. Recently organizations, such as the _The Nature Conservancy_ Organization have purchased and set aside considerable space for wild megafauna and plants. This would have been unthinkable 100 years ago. A modern concern for endangered species might protect them.

Figure 54 plots the GISP2 Central Greenland ice core surface air temperature record by (Alley 2004) through the period of interest. The GISP2 record is plotted, rather than the Vinther record, because the Vinther record only goes back to 11,700 BP and doesn't show the Younger Dryas.

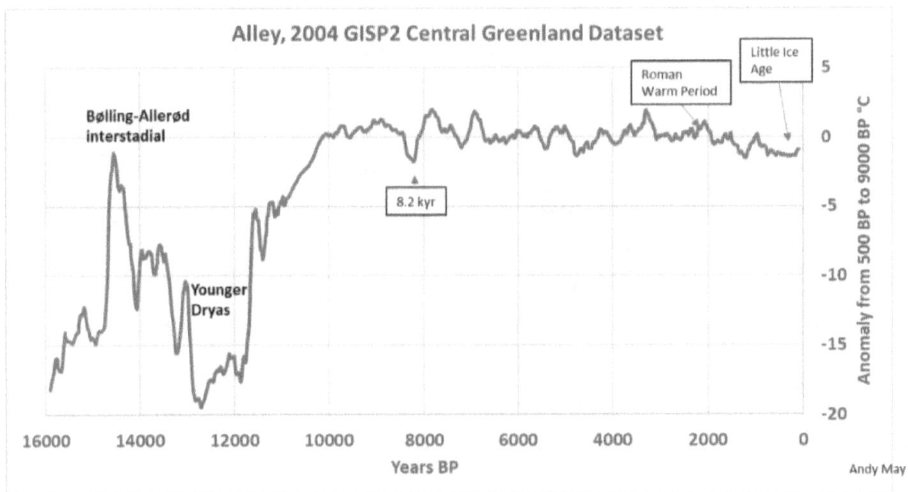

Figure 54. The Greenland Ice Sheet Project 2 (GISP2) temperature reconstruction, showing the Younger Dryas (Alley, 2004).

At least in Greenland, and probably over the whole Northern Hemisphere, it was very cold 12,000 years ago.

After the Younger Dryas the world warmed rapidly, especially in the Northern Hemisphere. The Central Greenland ice core reconstruction shows over 10°C (18°F) of warming in less than 144 years from 11,755 BP to 11,611 BP. The Vinther, et al. record starts during the temperature increase at 11,700 BP, where it shows a temperature of -8.4°C (anomaly from today) and a temperature anomaly of -3.22°C at 11,600BP. This is an increase of 5.2°C (9.4°F) in 100 years. The cool Younger Dryas period followed by abrupt warming put a lot of stress on all species, and undoubtedly some species went extinct. But, the species we especially notice in the fossil record are the larger creatures. Evidence that the surface air temperatures rose

143

5-10°C in just a few decades over the entire Northern Hemisphere is presented in (Severinghaus, et al. 1998). We have people concerned today because of a 0.9°C (1.6°F) rise in 137 years. Imagine an increase of ten times that in the past 140 years! This is what the Northern Hemisphere likely saw at the end of the Younger Dryas.

By 12,000 years ago, humans had spread throughout Africa, Eurasia and the Americas. The rise of civilization and the development of new hunting tools (bows and arrows, spears, and spear throwers – all invented more than 30,000 years ago) made Upper Paleolithic hunters very formidable. We know that Upper Paleolithic humans hunted megafauna 12,000 years ago and earlier and were present in areas where megafauna went extinct. In Australia, the megafauna disappeared within a few thousand years of man's arrival on the continent and in a time of stable climate. But elsewhere, like in Eurasia, the extinctions occurred during periods of dramatic climate change. It is likely that hunting by Upper Paleolithic humans and climate change both played a role in the megafauna extinction event. A summary of the geological evidence for both theories can be seen in (Barnosky 2008). Further, it seems clear that of the megafauna present 12,000 years ago, humans were the most adaptable to the changing world. To a large degree, humans simply outlasted and out-survived their competitors. Evolution at work.

Barnosky mentions that some megafauna are at risk of extinction today. Many of the large species have only survived in Africa. This is not a surprise; Africa sits on the equator and, except for the Mid-Holocene Transition, is generally less vulnerable to severe climate change (see Chapter 4). Because of their size megafauna require a lot of land to live in the wild. In the future this will restrict them to parks, private lands and zoos. But as man becomes more prosperous, he is more interested in preserving them and they are unlikely to extirpate. Upper Paleolithic humans were only interested in surviving. Humans today are more affluent, secure and will expend energy and resources to help species that we value survive. It is more likely that humans today will ensure the survival of megafauna, rather than threaten them.

Barnosky speculates that our current high level of large animal "biomass" is only being sustained by fossil fuels and is dominated by human "biomass." Strange way of putting it, but it might be true. His next prediction is that another large animal "biomass crash" is imminent because we are running out of fossil fuels. This is unlikely mainly because his prediction that we only have an 83-year supply of oil and gas is much too pessimistic. The paper states we have a 50-year supply of oil and a 200-year supply of gas. The gas supply is converted to oil-equivalent by using the <u>USGS conversion</u> of 6 MCF of natural gas to one barrel of oil equivalent.

Table 6, (May 2017i), gives a much more realistic estimate of known hydrocarbon resources that are technically recoverable. It is very conservative and does not include all known hydrocarbons. All the hydrocarbon resources in Table 6 can be produced economically at prices we have seen in the last 20 years, although some may not be economically viable at current (2017) prices. It appears that Barnosky was only counting "proven" or "probable" conventional oil and gas deposits, while most of the oil and gas known today are classified as "unconventional" (including oil shale and oil sand deposits). While the production of unconventional oil and gas is more expensive than conventional oil and gas, it is still an order of magnitude cheaper, in terms of energy returned on energy invested, than solar and wind. Nuclear and hydroelectric are even cheaper than coal or natural gas, as shown by (Weißbach, et al. 2013). The bottom line is that available energy supply is not the problem.

World Technically Recoverable Oil & Gas			
Summary	Oil & Gas	Supply	Supply	
	BBOE	days	years	Source
Proven	2,797.8	18,536.9	50.8	BP, 2016
Undiscovered conventional	1,666.3	11,039.8	30.2	USGS, 2012
Unconventional shale oil&gas	1,681.7	11,141.8	30.5	EIA
Oil shale (kerogen)	1,241.7	8,226.8	22.5	IEA
Oil sands & bitumen	781.4	5,177.1	14.2	IEA, ACS
Total	8,168.9	54,122.4	148.2	

Table 6. World technically recoverable oil and gas. Source (May 2017i).

While Upper Paleolithic (roughly 40,000 to 10,000 years ago) man played some role in the Quaternary megafauna extinction event, this was a very different human than we are (Holt and Formicola 2008). They were subsistence hunters in a very tough time. This caused physiological changes as they adapted to the tough and very cold climate. During this period, they had barely learned to farm, and even then, only in the most primitive way. When food walked by, they killed it and ate it. We have already seen that taking care of the environment is only possible when our income and security needs are taken care of (see Chapter 1, Figure 2). Today when a country's GDP in PPP$ (purchase power parity U.S. dollars) exceeds about PPP$2,000 per person, the environment improves rapidly.

Barnosky suggests that man's rise and other megafauna going extinct is simply a trade-off in biomass, thus as man's population grows more extinctions will occur among existing megafauna. I find this overly simplistic and highly speculative, considering our prosperity and our interest in preserving endangered species. There is a trade-off in available land, but biomass? With modern farming we can feed many more people and animals than in the past. We can also use energy to create a nearly infinite amount of clean water. Further, man has an obvious interest in preserving existing megafauna for esthetic and humanitarian reasons. They will not be able to run wild in a fenceless wilderness, but I doubt we will allow large animal species to go extinct. The best insurance for the survival of megafauna is a prosperous and secure human population.

Polar Bears

As Dr. Susan Crockford has reported in detail on her site polarbearscience.com, U.S. Fish and Wildlife Service (USFWS) researchers predicted that polar bears would die off to dangerously low levels if sea ice dropped below 3-5 million square kilometers on a regular basis (S. J. Crockford 2017b). This prediction was first made by the International Union for the Conservation of Nature (IUCN) in 2006. A second assessment was made by the U.S. Fish and Wildlife Service in 2008. Based on these predictions, polar bears were put on the vulnerable species list (Schliebe, et al. 2008).

The IUCN allows predictions of future populations to be used in requests to add species to their "red list" of vulnerable species. But, current rules for supporting data are much stricter than in 2008. It may very well be that polar bears would not be added to the list today using the same evidence (S. Crockford 2017).

Sea-ice extent has reached 3-5 million square kilometers on a regular basis since 2007, much earlier than expected, and this is long enough to assess the predictions (S. J. Crockford 2017b). The "critical" level of ice was reached many times, yet polar bears thrived. The population stayed stable or, perhaps, grew in number. The 2015 population of polar bears was estimated to be about 26,000. It has since been suggested that a low in the polar bear population was reached between 2004 and 2006 when the sea ice was thick and extensive (see Figure 55) (Bromaghin, et al. 2015). The population of bears in 2005 was estimated to be 22,500 (between 20,000 to 25,000). The population increased after the ice melted and polar bears were spending more time on land (Bromaghin, et al. 2015).

The loss of sea ice since 2007 has not affected the polar bear population, in the words of (Atwood, et al. 2016):

> "… no causal link between the patterns in polar bear vital rates and increased use of terrestrial habitat, …"

The authors suggest that behavior changes can ameliorate environmental changes. In the absence of ice, the bears simply moved onto land and did well. They then speculate that if the bears spend more time on land, due to the lack of ice, it may be a problem in the future. But, history has shown we should not make environmental decisions based upon speculation.

Once the lack of a link between sea ice extent and polar bear populations was established, Jeffrey Bromaghin, said in an interview on the paper (USGS 2014):

> "The low survival may have been caused by a combination of factors that could be difficult to unravel," said Bromaghin, "and why survival improved at the end of the study is unknown."

The population low point was 2004 to 2006, when the Arctic sea ice was extensive. The "survival of adults and cubs began to improve in 2007" when Arctic sea ice declined rapidly. This is exactly the opposite of what the environmentalists predicted. The September sea ice extent, the month with minimum Arctic sea ice, for the years discussed is shown in Figure 55.

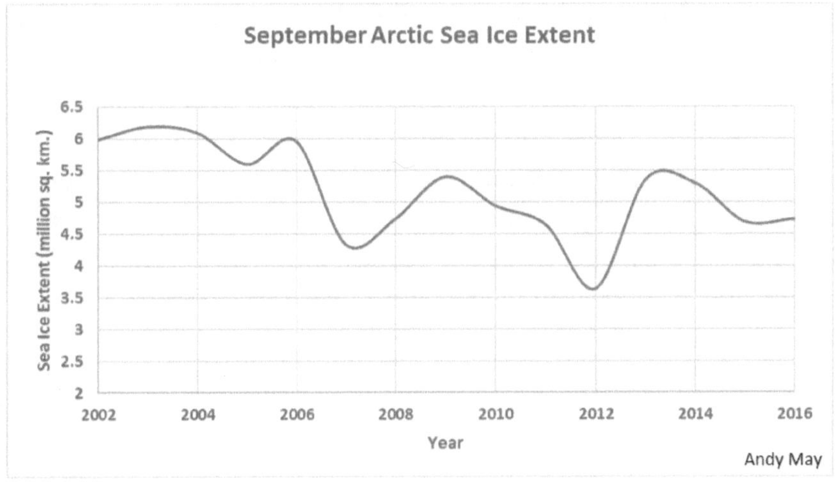

Figure 55. September Arctic Sea Ice Extent, Data from NSIDC (National Snow & Ice Data Center 2017).

The polar bear story is a good example of what can go wrong when one bases policy on environmentalist's predictions. In this case, the polar bear scientists concluded, erroneously, that polar bears needed sea ice to survive. Then they predicted if sea ice reached the 3-5 million square kilometer level or lower polar bears would suffer; this was also incorrect. Finally, they used a climate model to predict that the critical sea ice level would be reached in 2050 due to man-made climate change, which was also incorrect. It was reached a scant two years later and completely ignored by the bears, who simply lived on land in the absence of ice. Not a single prediction was correct, but the polar bear made the "vulnerable" list anyway.

Global Warming will shut down the Gulf stream and cause a mini-ice age.

This is a hypothetical scenario where melting Greenland ice dilutes the Atlantic between Greenland and Norway and stops the North Atlantic thermohaline circulation. This circulation is also called the "Atlantic conveyor" or the Atlantic meridional overturning circulation, abbreviated "AMOC." Based on a climate model, Michael Schlesinger has suggested that global warming could shut down the AMOC and cause catastrophic climate change (University Of Illinois At Urbana-Champaign 2004). Others have speculated that the AMOC is slowing down (Rahmstorf, et al. 2015).

Rahmstorf, et al. used climate models to identify a region of the North Atlantic that they assume varies from the Northern Hemisphere average temperature as a function of the strength of the AMOC. They use the temperature difference to create an "AMOC index." The index was never tested against actual measurements, but instead compared to a "state-of-the-art global climate model, the MPI-ESM-MR." Thus, a model was created that Rahmstorf, et al. assume represents the speed of the AMOC and tested by comparing it to another model.

The AMOC carries warm surface waters to the North Atlantic, where they cool, sink and are carried south by deep ocean currents. While it is true that if this circulation slows or stops it will trigger abrupt cooling, such as the 8.2 kyr event discussed in Chapter 4 (Li, et al. 2011), recent measurements by NASA show that the thermohaline circulation may actually have sped up slightly since 1993 (Buis 2010).

The NASA study was published in *Geophysical Research Letters* by Josh Willis (Willis 2010). Willis used Argo float (Argo 2017) data to build a 3D model of the AMOC from 2002 through 2009. The Argo floats measure the current speed and direction at 1,000, 1,500 and 2,000 meters. These measurements were used to build the model of the AMOC, which compares well with satellite altimetry measured sea surface height data. Sea surface height (SSH) varies with the AMOC because warmer water occupies more volume than cooler water.

He found no statistically significant change in the overturning circulation strength during the period from 2002 to 2009. He extended this period back to 1993 using the satellite altimetry SSH measurements, and the resulting graph (Figure 56) suggests that the AMOC has sped up slightly between 1993 and 2009. While a model was used in this study to compute the speed of the overturning circulation, the model was based upon actual measurements over an eight-year period.

Rahmstorf, et al. used little or no measurements at all to build their model or compute their estimate. They also used temperature reconstructions by Michael Mann (a co-author in the study), which should be greeted with suspicion, as discussed in Chapter 4. They merely assume that their "AMOC index" is related to the AMOC itself. Measurements of the AMOC were available when Rahmstorf, et al. was written, and they are mentioned in the text, but only a very incomplete set of points are shown in their Figure 5 on page 478 of the paper (Rahmstorf, et al. 2015).

Rahmstorf, et al. do mention the study by Josh Willis discussed above, but only to say the two studies are consistent in what they call the recovery period for the AMOC index from 1990 to the present day. Their study shows the AMOC slowing from 1930 to 1990 and then increasing in velocity until 2010. Using Mann's reconstructions, they extend their AMOC index back to 1000AD, but this is speculative.

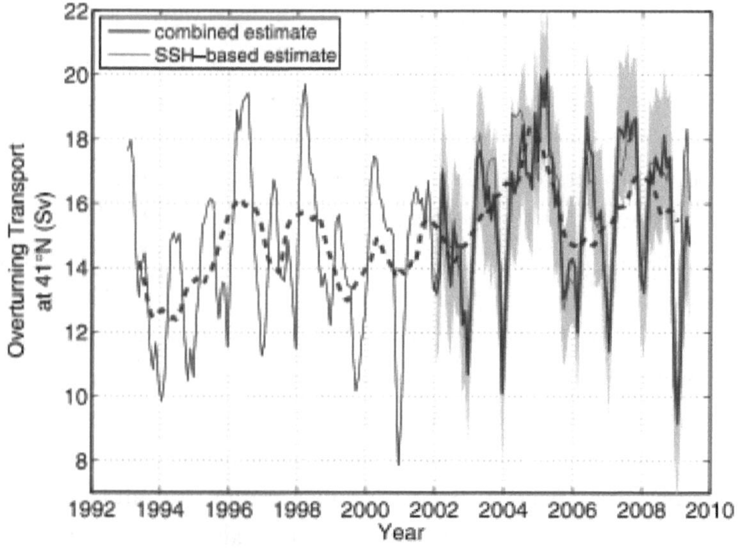

Figure 56. AMOC speed, showing a slight increase from the 1990s to the 2000s. From (Willis 2010) in Geophysical Research Letters, used with permission.

Figure 56 shows Willis' estimate of the overturning circulation transport speed from 1993 to 2009. From 2002 to 2009 data from satellite altimetry and from Argo float measurements are used to measure the speed, shown with a blue line. The blue shading indicates error bounds. Over this period, we see no significant change in speed.

The period from 1993 to 2002 is estimated with a satellite altimetry SSH model of the speed. This is shown with the thin, solid, red line. Some increase in speed is suggested using the model. The dashed, red line is a one-year moving average of the satellite altimetry-based velocity estimate.

8,200 years ago (more precisely 8.31 kyr to 8.18 kyr), an enormous fresh water lake (Lake Agassiz-Ojibway), in the area along the border between Canada and the U.S., broke through an ice dam that was left over from the most recent glacial advance and flooded the North Atlantic with fresh water. This raised sea level 0.8 to 2.2 meters (Li, et al. 2011). It also lowered the salinity of the Atlantic surface water, halted the thermohaline circulation in the North Atlantic and cooled the planet for hundreds of years (Michalek 2017) (Li, et al. 2011). We often simply refer to one cooling, the 8,200 BP cooling period, but, there were two, one was 8,490 years ago and the other from 8,340 BP to 8,180 BP. Per Lomborg's *Cool It:*

"The Gulf Stream last shut down some 8,200 years ago, when the final glacial ice sheets in North America melted and a giant pool of freshwater built up around the area of the Great Lakes. One day, the ice dam broke, and an unprecedented amount of freshwater flooded the North Atlantic and disrupted the sinking salty water from the Atlantic conveyor. This pushed Europe into a little ice age for almost one thousand years. …

Yet the relevance of such a story [today] crucially depends on the Greenland melt being on the same order of magnitude as the ancient freshwater pool—and it is not. Over the coming century, the IPCC expects Greenland to melt almost one thousand times less than what happened 8,200 years ago. A team of modelers looked at what would happen if Greenland melted at triple the rate expected by the IPCC—or, as they put it, at the "upper limit on possible melting rates." Although they see a reduction in the Gulf Stream, they find "its overall characteristic is not changed" and that "abrupt climate change initiated by Greenland ice sheet melting is not a realistic scenario for the 21st century. …

This is also why the IPCC, in its 2007 report, is very clear about the Gulf Stream: 'None of the current models simulates an abrupt reduction or shut-down.' The IPCC's models expect somewhere from no change to a Gulf Stream reduction of 50 percent over the coming century, but no models show a complete shutdown." Lomborg, Bjorn. *Cool It* (Kindle Locations 1334-1394).

Josh Willis comments on the possible effect of Greenland's ice sheet melting on the AMOC in the introduction to (Willis 2010):

"Although acceleration in the rate of ice loss from Greenland has already been well documented [(Rignot and Kanagaratnam 2006) (Wouters, Chambers and Schrama 2008)], model results suggest that the current ice loss rates are too small to slow the AMOC [(Hu, et al. 2009)]."

This idea was the inspiration for the 2004 film *The Day after Tomorrow* starring Dennis Quaid and directed and written by Roland Emmerich. This very imaginative film is filled with gross scientific inaccuracies but did reasonably well at the box office. Critics panned it, as did most

Earth scientists I know who bothered to watch it. Hollywood is not a source of accurate information on climate change.

Chapter 5 Conclusions

Bottom line, there is no discernable trend in extinctions, up or down. We are certainly not in a "major extinction," "mass extinction," or "great extinction" event. The Gulf Stream and the world's thermohaline circulation system are in fine shape, and there is no "Lake Agassiz-Ojibway" waiting to spill into the North Atlantic to shut them down. Ice on Greenland cannot melt fast enough to affect the North Atlantic surface salinity, to the degree necessary, to repeat the 8,200 BP cold spell.

The Quaternary megafauna extinction was very sad, and both Upper Paleolithic humans and radical natural climate change played a role in those extinctions according to the data we have available. However, comparing this very minor extinction event to the five great (or "mass") extinctions, as Barnosky and Kolbert attempt to do, betrays a complete lack of proportion and an ignorance of the geological past. The true great extinctions were horrific events (Brannen 2017), and the loss of 178 species of large animals is several orders of magnitude too small to qualify. For this event to be classified as a great or mass extinction event it would require the extinction of over half a million species, and nothing like that is happening today or in the foreseeable future.

Dr. Susan Crockford has documented the saga of polar bears being erroneously declared "vulnerable" based on "expert" predictions quite well, and there are lessons to be learned from this fiasco. One cannot base public policy on unvalidated models and predictions. This recent trend of believing, without question, model results; or worse using model results as if they were data, needs to stop. At some point it will lead us off a cliff.

Chapter 5 Summary

Great (or "Mass") extinctions are horrific events, and the Holocene extinction data that we have available does not suggest we are in one, or even close. As man became farmers and ranchers, starting about 12,000 years ago, and has taken over more land area for these pursuits, the land area available for large mammals, especially predators, has decreased. This has decreased these populations.

Species are protected gene pools and are isolated to some degree from other species. When a species becomes extinct, its gene pool disappears. In fact, extinction can only technically occur for a species or higher taxa. The media, and some government announcements, often talk about sub-species or populations going "extinct." An example, is the so-called "extinction" of the eastern cougar. The eastern cougar is a population of the sub-species North American cougar, which is a sub-species of cougar. Restricting the range of the North American cougar to the western U.S. is not an extinction. The gene pool survives.

Human hunting and rapid natural climate change during, and just before, the Holocene has caused decreases in some mammal populations, as well as some extinctions. However, the number of known extinctions is very small relative to the millions of known species in the world, and there is no cause for alarm. These extinctions are sad, but they are not unusual in either their rate or their number and do not exceed the accepted average rate of species extinctions of 1 to 2 per year.

Further, man is very different today than 12,000 years ago. As we have become more affluent we can afford to help endangered species we care about, and we do. The climate 12,000 years ago was horrible (see Figure 54). Humans were in survival mode; they needed to eat and certainly didn't care about animals except as food. This is very understandable, and attempting to tie that time to this one through a simple "biomass" exchange, as Barnosky tries to do, is an error in logic.

Polar bears are not endangered, and their population appears to be increasing. They were added to the vulnerable species list based upon "expert" environmentalist projections that turned out to be wrong. This was done even when it was clear that no actual data supported the "vulnerable" classification. No models or projections should be used for species classifications like this unless actual data supports the projections.

Model projections and "expert" opinions that melting of the Greenland ice sheet will shut down the Atlantic meridional overturning circulation (AMOC), causing catastrophic cooling, are at odds with measurements made by NASA. Even at the fastest possible Greenland melt rate, there simply is not enough melt water available in Greenland for this to happen. It's just science fiction. In the next chapter we consider the effect of climate change and global warming on mortality and disease.

CHAPTER 6

Climate-Related Deaths and Insecurity

In this chapter we will discuss the assertion that there will be more climate-related deaths due to man-made global warming. This assertion is complicated by the fact that currently more people die during cold weather than hot weather. Further, identifying the precise cause of a death can be difficult, especially when there are multiple apparent causes. For example, if a car accident is the cause of death and the roads are icy, did the accident occur primarily because of the icy road, or were other factors more important?

There will be more heat-related deaths

The IPCC AR5 report does not have much to say regarding climate-related mortality. They do mention that heat-related deaths will increase in several places; the following is from page 49 of the WG2 technical summary (Field, et al. 2014):

> "At present the worldwide burden of human ill-health from climate change is relatively small compared with effects of other stressors and is not well quantified. However, there has been increased heat-related mortality and decreased cold-related mortality in some regions as a result of warming (medium confidence)."

In 2014, a National Health Statistics Report "Deaths Attributed to Heat, Cold, and Other Weather Events in the United States, 2006-2010" was published (Berko, et al. 2014). The report used death certificate data collected by the <u>CDC</u> (U.S. Centers for Disease Control and Prevention). Over the five years of the study, 10,649 people died from weather-related causes, 31 percent from excessive heat, 63 percent from excessive cold and 6 percent from other weather-related causes like floods or lightning.

In a very interesting paper that compares different mortality databases, P. Dixon and colleagues from the Office of Climatology in the Department of Geography at Arizona State University report that the CDC data from 1979 to 1999 found that there were 3,829 heat-related deaths due to weather conditions. Over the same period there were 15,707 deaths due to hypothermia or excessive natural cold, excluding anthropogenic cold deaths (Dixon, et al. 2005). Examples of anthropogenic heat- and cold-related deaths are boiler room accidents, kitchen accidents and factory accidents involving excessive heat or cold. In examining several studies of heat and cold-related mortality, Dixon, et al. report:

> "Interestingly, depending on the database used and the compiling U.S. agency, completely different results can be obtained. Several studies show that heat-related deaths outnumber cold-related deaths, while other studies conclude the exact opposite. We are not suggesting that any particular study is consistently inferior to another, but, rather, that it is absolutely critical to identify the exact data source, as well as the benefits and limitations of the database, used in these studies."

In addition to the large difference between heat-related deaths and cold-related deaths, deaths in general are lower in the summer than in the winter. Figure 57 plots the gross mortality for the U.S. by month. Optimal mortality (meaning the fewest deaths) is always in the summer. Yet, some studies that show heat-related deaths are more numerous than cold-related deaths "detrend" the gross mortality data by taking out the pattern in Figure 57 and working with the residuals (<u>Kalkstein, 1991</u>). This takes out much of the effect of winter on mortality and is a questionable methodology, in my opinion, for our purpose.

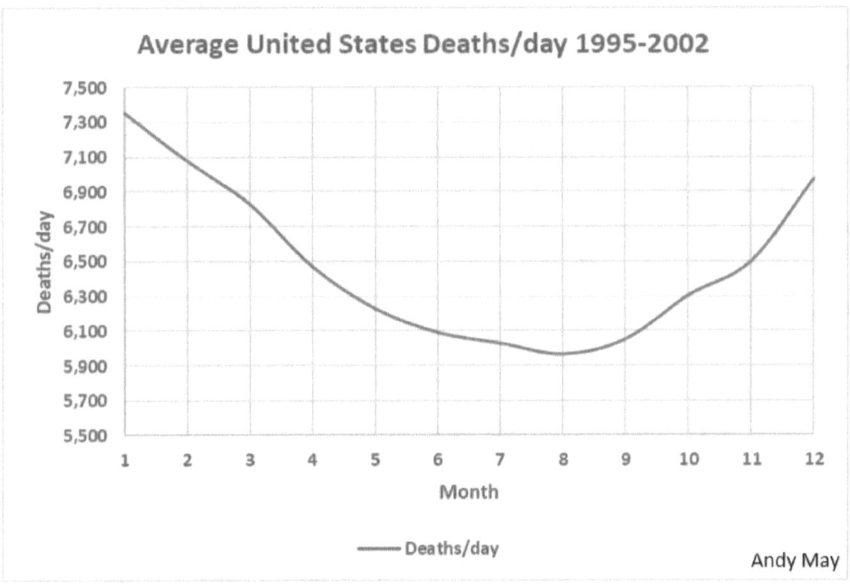

Figure 57. The percent of all deaths by month for the U.S., data source ABC News.

Detrending the gross mortality data is useful for examining the effect of a specific weather event on mortality. The residuals from the trend could be used to investigate the deaths due to a heat wave or snow storm, for example. But, we are not interested in weather events in this chapter. We are interested in the effect of long-term climate changes on mortality. Thus, the trend is more useful to us than the residuals.

Another source of weather-related mortality statistics is the <u>NCDC</u> <u>"Storm Data"</u> dataset (NOAA 2017d). Dixon, et al., 2005 provides a very thorough examination of this database and concludes:

> "Temperature extreme deaths listed in NCDC's Storm Data are skewed heavily toward heat-related deaths."

Because the database is based on media reports, it is also skewed toward media reported storm events and misses individual weather-related deaths that occur in the absence of a storm. These occur more often in cold winter weather, thus creating a warm bias. Further, media reports during an event often contain errors, and these are not corrected later in this database. For whatever reason, the NCDC storm database seems to overestimate heat-related deaths and underestimate winter-related deaths. See (Dixon, et al. 2005) for a complete discussion.

The CDC data has problems as well but is far more comprehensive and should be more reliable, in our opinion. Classification of the cause of death is mostly done by professionals and it is quality controlled, but misclassifications do occur. However, it seems likely that weather-related cold deaths outnumber weather-related heat deaths in the U.S. by a factor of two or more, as shown in the CDC data. Dixon, et al., 2005 do not agree and conclude:

> "Of the datasets identified in this study, the one that appears to be least influenced by the above limitations [misclassifications of cause of death] is gross mortality. However, the gross mortality data must be detrended in order to remove a persistent winter-dominant death maximum."

Removing "a persistent winter-dominant death maximum" prior to the analysis of weather-related deaths introduces a bias and suffers from the fallacy of removing the forest to study a tree. It's OK if the tree is all you want to study, but not so useful if you want to study the forest.

If the Earth warms there will certainly be more heat-related deaths. But, warmer temperatures will reduce the number of cold-related deaths. Since, currently there are more deaths due to cold, we expect the avoided cold-related deaths to outnumber the increase in heat-related deaths. For the foreseeable future, global warming will save lives.

"The first complete survey for the world was published in 2006, and what it shows us very clearly is that climate change will not cause massive disruptions or huge death tolls. Actually, the direct impact of climate change in 2050 will mean fewer dead, and not by a small amount. In total, about 1.4 million people will be saved each year, due to more than 1.7 million fewer deaths from cardiovascular diseases and 365,000 more deaths from respiratory disorders. This holds true for the United States and Europe (each with about 175,000 saved), as for the rest of the industrialized world. But even China and India will see more than 720,000 saved each year, with deaths avoided outweighing extra deaths nine to one." Lomborg, Bjorn. *Cool It* (Kindle Locations 681-686).

The 2006 paper that Lomborg refers to is (Bosello, Roso and Tol 2005) "<u>Economy-Wide Estimates of the Implications of Climate Change: Human Health</u>". The paper computes the climate-related health effects of a 1.16°C global average temperature change in 2050. Figure 58 shows where this temperature falls on the IPCC AR5 temperature prediction scenarios. It is roughly the temperature change associated with the RCP 4.5 scenario.

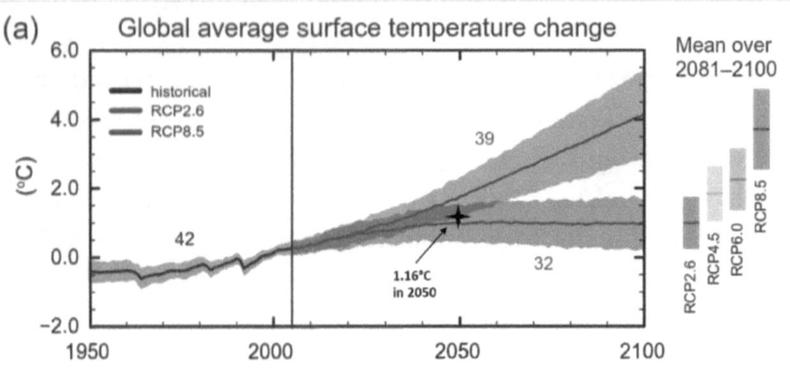

Figure 58. Temperature projections to 2100 for two Representative Concentration Pathways (RCP). Source IPCC AR5 The Physical Science Basis, Summary for Policymakers, page 21, SPM.7a, used with permission (IPCC 2013).

"For almost every location in the world, there is an "optimal" temperature at which deaths are the lowest. On either side of this temperature—both when it gets colder and warmer—death rates increase. However, what the optimal temperature is, is a different issue. If you live in Helsinki, your optimal temperature is about 59°F, whereas in Athens you do best at 75°F. The important point to notice is that the best temperature is typically very similar to the average summer temperature. Thus, the actual temperature will only rarely go above the optimal temperature, but very often it will be below. In Helsinki, the optimal temperature is typically exceeded only 18 days per year, whereas it is below that temperature a full 312 days. Research shows that although 298 extra people die each year from it being too hot in Helsinki, some 1,655 people die from it being too cold."

164

"It may not be so surprising that cold kills in Finland, but the same holds true in Athens. Even though absolute temperatures of course are much higher in Athens than in Helsinki, temperatures still run higher than the optimum one only 63 days per year, whereas 251 days are below it. Again, the death toll from excess heat in Athens is 1,376 people each year, whereas the death toll from excess cold is 7,852."

"This trail of statistics leads us to two conclusions. First, we are very adaptable creatures. We live well both at 59°F and 75°F. We can adapt to both cold and heat." Lomborg, Bjorn. *Cool It* (Kindle Locations 344-355).

The IPCC predicts that the global temperature in 2100 will be about 2°C higher than today. But nobody lives at the global average. Temperatures are rising more in Siberia and Canada than anywhere else, and Antarctica is getting colder. Land temperatures rise faster than ocean temperatures, and temperatures at night are rising faster than daytime temperatures. The reality of climate change isn't necessarily bad; it doesn't mean a fierce heat wave. The temperature changes we are considering are small and will be welcomed by many.

Lomborg estimates in *Cool It* that for the temperature increase of about 0.4°C observed from the 1970s to 2005, we get about 620,000 avoided cold deaths and 130,000 extra heat deaths. Thus, global warming (1970-2005) saved almost five times more people than it killed.

According to the <u>Fourth National Climate Assessment</u> (Wuebbles, et al. 2017) extreme cold events and cold waves are less frequent. They also say, rather disingenuously, that heat waves "have become more frequent in the United States since the 1960s." This is misleading because the 1960s were unusually cool (see Figure 59) and because the most severe heat waves in the U.S. occurred in the 1930s according to the (EPA 2017b), as seen in Figure 60.

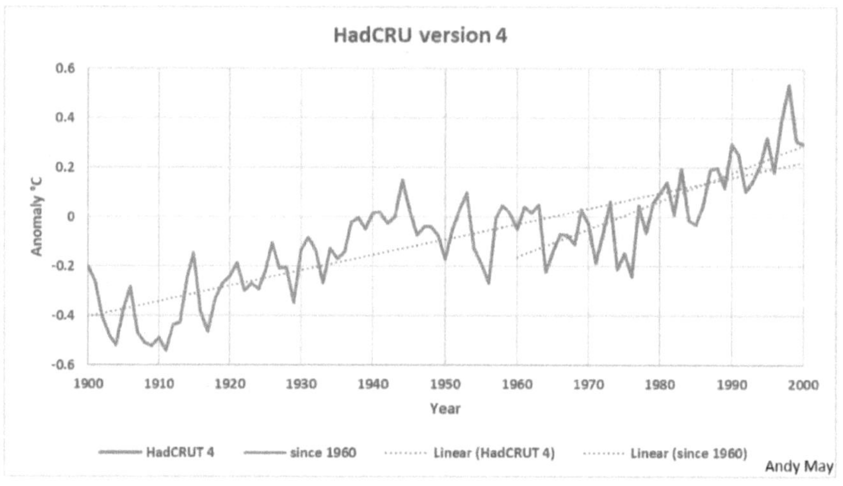

Figure 59. HADCRUT4 global temperature reconstruction. Data: HADCRUT 4, (Met Office Hadley Centre and Climatic Research Unit 2017.

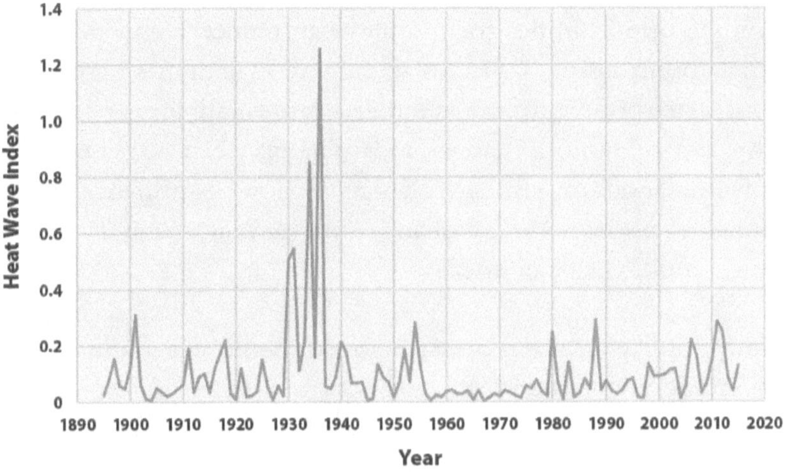

Figure 60. The EPA heat wave index 1890-2015, source (EPA 2017b). Used with permission.

Insecurity

The IPCC WGII AR5 Technical Summary claims that climate change will cause great uncertainty and increase population insecurity. Most of the claims in the section are easily dismissed as silly, like the threat of war will be increased, or cultural values will be eroded. Some have attributed the Syrian civil war and the European refugee crisis to climate change (Arsenault 2015). But other more serious claims are worth discussing. For example:

"Climate change over the 21st century is projected to increase displacement of people (medium evidence, high agreement). Displacement risk increases when populations that lack the resources for [a] planned migration, experience higher exposure to extreme weather events, in both rural and urban areas, particularly in developing countries with low income. Expanding opportunities for mobility can reduce vulnerability for such populations. Changes in migration patterns can be responses to both extreme weather events and longer-term climate variability and change, and migration can also be an effective adaptation strategy. There is low confidence in quantitative projections of changes in mobility, due to its complex, multi-causal nature."

Warming, unlike war or repressive governments, is a gradual thing, and air conditioning is currently not very expensive if electricity is available and cheap. If an area becomes too warm, people may want to move to cooler places, but we are not talking about rapid warming. Warming is expected to be at a rate of 0.1° to 0.2°C per decade. This is also a time of rapid increases in standard of living. As people become more affluent, they will have many more options to adapt to the changes in-place or by moving to a cooler locale. "Wealthier is Healthier" as noted by (Pritchett and Summers 1996).

The IPCC also claims that climate change will slow down economic growth and make poverty reduction more difficult, increase hunger, etc. But, we have examined these issues elsewhere in this book and shown that there is no evidence, other than unvalidated models, to support these projections.

Generally, the idea of insecurity caused by climate change is based upon the effect of severe weather on poor populations that cannot build the infrastructure necessary to protect themselves (Busby, et al. 2013). In studies that attempt to rank countries by vulnerability to climate change, like (Busby, Cook, et al. 2014), the poorer countries rank higher than the wealthier countries. Are the countries vulnerable to climate change or poverty and corruption? Should we tackle their insecurity by reducing our use of fossil fuels, hoping to change the climate? Countries that have lifted themselves out of poverty, like China and India, have done so by greatly increasing their use of fossil fuels to supply their populations with energy (Pachauri and Jiang 2008). If the problem in these "climate insecure" countries is poverty and not climate, are we making their situation worse?

Pachauri and Jiang (2008) examined the rise in fossil fuel energy use in China and India and the impact on the populations. They write:

"It is clear though that the transition from solid fuels [mainly indoor use of wood, dung and coal] to cleaner and more efficient fossil-based energy sources has significant social benefits and leads to an improvement in the local and indoor environment. A shift away from solid fuels is associated with improvements in living conditions and health, particularly for poor women and children most affected by indoor air quality."

The switch from burning wood and other biomass to fossil fuels has also resulted in greater literacy and more entrepreneurship. The former by allowing children more time and light to study and the latter by providing the transportation and power required to start new businesses. One cannot see the insecurity of a population only through the lens of climate change and only consider mitigation through the reduction of CO_2 emissions as a solution. Curtailing CO_2 drastically reduces a populations prosperity and their ability to adapt; the treatment can be worse than the disease.

Chapter 6 Conclusions

We have shown that, according to the CDC death certificate database, more lives will be saved due to lower cold weather mortality than caused by higher temperatures in the U.S. In addition, the World Health Organization has found that studying climate-related mortality is very complex, and the results of their study are uncertain. They conclude:

> "Climate change is expected to affect the distribution of deaths from the direct physiological effects of exposure to high or low temperatures (i.e. reduced mortality in winter, especially in high latitude countries, but increases in summer mortality, especially in low latitudes). However, the overall global effect on mortality is likely to be more or less neutral. The effect on the total burden of disease has not been estimated, as it is unclear to what extent deaths in heat extremes are simply advancing deaths that would have occurred soon in any case."

The U.S. and many other developed countries in the temperate latitudes have good statistics on the cause of death, and in these areas, cold weather-related deaths outnumber hot weather-related deaths. Thus, in these areas, warming will reduce the weather-related deaths overall. As shown in Dixon, et al. many more deaths occur in the winter than in the summer in any case, regardless of the cause. However, this is in temperate areas, and 50 percent of the surface of the Earth lies between 30°N and 30°S. The data for this latitude band is incomplete and not as accurate. So, it seems likely that we should accept the WHO conclusion that we simply do not know what the global effect of global warming will be over the entire world, but it will likely be "more or less neutral."

Climate changes, whether natural or man-made, are very likely to make some populations insecure. Droughts, flooding, severe weather and other weather-related problems will increase in some areas and other areas will benefit. Due to these changes, people will move, as they always have in such circumstances. The cold period from about 100 AD to 400AD, near the end of the Roman Empire, may have initiated the so-called "Migration Period" (Schrover 2008). This resulted in the Visigoths, Huns and other migrating people conquering large parts of the Western Roman Empire. The timeline in our post on "Climate and Civilization in the last 4,000 years" (May 2016f) marks the cold period on the Central Greenland temperature reconstruction. You can also see the dip in temperatures at 100AD in Figure 52.

Each city on the Earth has an ideal temperature, as Lomborg has pointed out. This ideal temperature is usually close to the average local temperature in the summer. Above this temperature and below this temperature deaths increase. The range of ideal temperatures is large, much larger than the average warming expected in the various climate model projections for the next several hundred years. Thus, as man has spread over the planet he has already adapted to more extreme temperatures than we are likely to see anywhere in the next few hundred years. In Helsinki, where the ideal temperature is 59°F, people do well. They also do well in Athens where the ideal temperature is 75°F and in Philadelphia where they do well at 80°F. It does not seem that climate change is a health risk.

Chapter 6 Summary

Claims that climate change will increase mortality are not supported by the data available. Mortality, in temperate climates, is always higher in the winter and lower in the summer. At least one worldwide study (Bosello, Roso and Tol 2005) concluded that 1.4 million people will be saved each year due to global warming. But, the data for most of the world is sparse, so the conclusion we have accepted, from World Health Organization, is that global warming and climate change will probably have no overall effect on mortality.

Climate changes will occur over the world, whether natural or man-made, and this will make some people insecure. Some of the affected people will move to places where the climate is better, and some will adapt to the changes. Adaptation is facilitated by technology and cheap, abundant energy. More people will have air conditioning and heating (other than burning dirty biomass, wood or coal indoors) available as the world becomes more affluent. In a similar manner, more people will have the option of moving to better climates as they become more affluent. The one thing we do not want to do is make energy more expensive or make people poorer. This reduces their ability to adapt.

If global warming doesn't increase disease or mortality, does it increase extreme weather? We deal with these claims in the next chapter.

CHAPTER 7

Global Warming and Extreme Weather

In this chapter we will discuss the connection, if any, between climate change and extreme weather. In the <u>IPCC WGII AR5 Technical Summary</u>, page 52, they list the following risks of climate change, among others (Field, et al. 2014):

> "Virtually certain that, in most places, there will be more hot and fewer cold temperature extremes as global mean temperatures increase, for events defined as extremes on both daily and seasonal time scales."

Global warming and the frequency and severity of natural disasters

The cost of natural disasters, including hurricanes, has been increasing with time. Most of the reason is more people live on the coast or in areas prone to natural disasters (forests, flood plains, mountain sides, next to levees, etc.), and they build very expensive buildings in these places. The IPCC acknowledges this, but in a slightly devious way, in <u>IPCC WGII AR5 Technical Summary</u>, page 49 (Field, et al. 2014):

> "Economic losses due to extreme weather events have increased globally, mostly due to increase in wealth and exposure, with a possible influence of climate change (low confidence in attribution to climate change)."

This statement is technically correct. There is "low confidence" in a *possible* influence on extreme weather by climate change. But, it is worded cleverly, so that one can easily read into it that man-made climate change has some influence on extreme weather, and many have.

Dr. Roger Pielke Jr. does not believe reducing fossil fuel use will influence extreme weather, although he does believe fossil fuel use should be curtailed for other reasons (an opinion we do not share).

"Adaptation and mitigation are not trade-offs but complements that address different issues on very different timescales of costs and benefits. If a policy goal is to reduce the future impacts of climate on society, then energy [fossil fuel] policies are insufficient, and indeed largely irrelevant, to achieving that goal. There are other sensible reasons for efforts to accelerate decarbonization; protecting us from disasters is not one of them, and arguments and advocacy to the contrary are not in concert with research in this area. Governments and businesses are already heavily invested in climate policy and thus should focus resources on decisions likely to be effective with respect to policy goals. In the context of extreme events, such decisions might focus increasingly on land use, insurance, engineering, warnings and forecasts, risk assessments, and so on." Pielke Jr., Roger. *The Climate Fix: What Scientists and Politicians Won't Tell You About Global Warming* (p. 189).

Thus, the focus on planetary CO_2 reduction to mitigate planetary extreme weather is pointless. Instead, focus on local risks, properly price disaster insurance, discourage people from building in high risk areas, build protective infrastructure and so on. Be aware of the risks in your own community, and deal with them locally. He follows with this:

"… poor countries, lack a basic resilience in the face of climate extremes, there is much work to be done to improve adaptive capacities. Such policies make sense independent of human-caused climate change, but they will also make these communities more robust in the face of human-caused climate change." Pielke Jr., Roger. *The Climate Fix: What Scientists and Politicians Won't Tell You About Global Warming* (p. 190).

In other words, climate disasters are much worse in poor communities. Indeed, all disasters are. Poor communities are much less adaptable than wealthy, developed communities. Wealthy countries are also healthier (Pritchett and Summers 1996) and have cleaner environments (see Figure 2 in Chapter 1). Fossil fuel use is what makes <u>developed countries wealthy</u> (May 2017b); taking it away will make them poorer and more vulnerable to natural disasters.

There is another reason the world will not give up on fossil fuels until something comes along that is both cheaper and better. The cost of energy is a <u>fundamental component</u> of standard of living (May 2017b). Pielke, Jr. describes the "iron law of climate policy." This unbreakable law is that all countries expect economic growth. People in wealthier countries will pay some amount for environmental goals because they want clean air and water. But, they will not sacrifice their prosperity for it, and they will not accept a declining economy.

I find Pielke Jr.'s book *The Climate Fix* a little frustrating. He often says, in various ways:

"Make no mistake: carbon dioxide matters a great deal." Pielke Jr., Roger. *The Climate Fix: What Scientists and Politicians Won't Tell You About Global Warming* (p. 18).

The book destroys the connection between CO_2 and climate extremes, the iconic "2°C limit" on warming, as well as any positive effect of reducing CO_2 emissions. He also shows that there is no evidence that a "tipping point" temperature exists. At least in this book, he never really explains why CO_2 emissions need to be reduced or why current global warming is a bad thing. My question is why? We don't need to stop using fossil fuels to have clean air, as we have found in the U.S. and Europe. We can have both. In fact, properly used fossil fuels are safer than <u>burning wood</u> (May 2017j) (Energy Policy Institute at the University of Chicago 2017) in a fireplace or stove. This quote is telling:

"John Beddington, science adviser to the UK government, said ... in early 2010: 'It's unchallengeable that CO_2 traps heat and warms the Earth and that burning fossil fuels shoves billions of tonnes of CO_2 into the atmosphere. But where you can get challenges is on the speed of change. When you get into large-scale climate modeling there are quite substantial uncertainties. On the rate of change and the local effects, there are uncertainties both in terms of empirical evidence and the climate models themselves.' Even with uncertainties about the future, there is ample evidence, broadly accepted, that humans are influencing the global earth system. Such influences carry with them a risk of undesirable outcomes." Pielke Jr., Roger. *The Climate Fix: What Scientists and Politicians Won't Tell You About Global Warming* (p. 32).

He seems to be saying humans are putting billions of tons of CO_2 in the air, and we have no idea what the effect is going to be, but since man is doing it; it must be bad, and we need to stop it. We hear this sort of twisted logic all the time. We know the planet has been warming since the Little Ice Age, the coldest period in the Holocene. Could this be a natural reversion to the mean? This is not the first book or article

I've read on climate science where the body of the presented evidence is at odds with the conclusions.

Another telling quote from the book:

> "The literature on this topic is so vast that one could easily cherry-pick a few studies suggesting that the impacts [of CO_2] may be benign or, in contrast, that those impacts may be catastrophic. Science cannot presently adjudicate between these possibilities, or even give reliable odds on particular outcomes, …. Many, if not most, scientists believe that the impacts will be on balance negative and significant." Pielke Jr., Roger. *The Climate Fix: What Scientists and Politicians Won't Tell You About Global Warming* (p. 61).

We don't know whether CO_2 is good or bad, nor can we measure the impact, but the "gut feeling" of most scientists is it will be bad. *Really?* Any investor will tell you that investing "from the gut" is a sure way to lose money. Is it a good company or sector? A fair price? What is the dividend? What is the expected total return? You need facts to invest wisely. Think of investing in reducing CO_2 emissions like you would any smart investment. Compare the investment to alternatives. Our point is, there are an array of solutions out there for extreme weather events and for climate change. We need to know which one has the best rate-of-return. Which one does the best job at the cheapest cost?

The most important part of assessing risk is to be quantitative. What is the risk and what is the cost of a bad outcome? What is the earliest time a bad outcome could occur? What options do we have to reduce the risk? How expensive is each option, and when do we have to pay for it? Don't pay for a solution until you know you have a problem and how bad the problem is or will be.

Global warming will increase the frequency and severity of hurricanes

There is a recent climate model-based study (Emanuel 2017) that attempts to quantify the probability of a Hurricane Harvey type of rain event, in the Houston area, in the period 1981-2000 by analyzing retrospective meteorological forecast data from the U.S. National Center for Atmospheric Research, NASA and the European Center for Medium Range Weather Forecasts (ECMWF). He then uses the IPCC RCP8.5 scenario results from six global climate models for the same area to predict a Hurricane Harvey probability for the period 2081-2100. Finally, he assumes that the change in probability is linear during the 21st century. He calculates that a Hurricane Harvey event is a 1 in 100-year event in 2000 and a 1 in 5.5-year event in 2100. This makes the probability of the event, in 2017, 1 in 16 years. His conclusions are based entirely on computer models and the RCP8.5 scenario, although actual data may have made some difference in the NOAA and European weather forecast re-analyses. The probabilities are only valid if the model is accurate. And, as we have seen the models have not been validated, thus it is a speculative exercise of little practical value (Curry 2017).

Roger Pielke Jr. has pointed us to a paper (Ritchie and Dowlatabadi 2017) that concludes the RCP8.5 scenario is implausible:

"Accounting for this bias [toward the use of coal] indicates RCP8.5 and other 'business-*as*-usual scenarios' consistent with high CO_2 forcing from vast future coal combustion are exceptionally unlikely. Therefore, SSP5-RCP8.5 should not be a priority for future scientific research or a benchmark for policy studies." (Ritchie and Dowlatabadi 2017)

Thus, Emanuel's calculation of the odds of Harvey occurring in 1981-2000 may be correct, but his projection from that time into the future is very unlikely to be correct.

(Risser and Wehner 2017), in *Geophysical Research Letters*, is a statistical study that attempts to model ENSO (La Niña and El Niño) precipitation data from the Global Historical Climatology Network, or GHCN. The stations used are shown in the left-hand portion of Figure 61. The gridded and contoured precipitation data is shown in the middle map. The NOAA Advanced Hydrologic Prediction Service (AHPS) estimated precipitation, based on radar and rain gauge data, is presented in the right-hand map in Figure 61. The study finds that human-induced climate change likely increased the chances of the observed Harvey precipitation by a factor of 3.5. They claim that man-made climate change increased the Harvey rainfall amount by at least 19 percent.

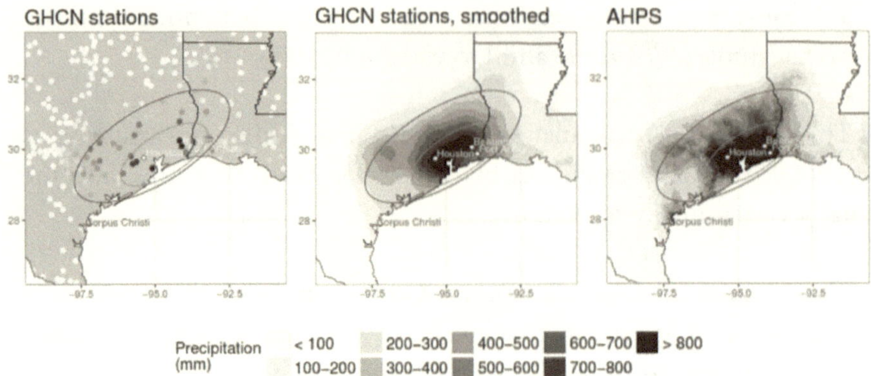

Figure 61. Global Historical Climatology Network stations in the Houston, Texas, area. Source (Risser and Wehner 2017), used with permission.

The model assumes that specific humidity increases, according to the <u>Clausius-Clapeyron</u> relationship, by 6-7 percent per degree of local warming in the absence of dynamical (wind) changes. They expect extreme precipitation to increase by the same amount.

Finally, they assume that the only influences on temperature are the atmospheric carbon dioxide concentration and ENSO (the Niño3.4 index to be specific). The modeling was done for the Houston area alone. They conclude that:

"[A 316 mm day rainfall] total has become much more commonplace: from a several hundred-year storm in 1950 to a 25-50-year storm in 2017. Our covariate-based analysis indicates that this change is due to the anthropogenic increase in atmospheric CO_2 concentrations and not natural ENSO variability."

While I applaud Risser and Wehner for using actual data in their study, I find fault with their assumption that temperature is the only driver for precipitation and that the Clausius-Clapeyron relationship (see below for a description of this relationship) is the only driver for specific humidity. These assumptions are speculative. Finally, assuming that the only drivers for temperature are CO_2 and ENSO is extremely speculative. Other natural drivers for both global and local temperatures are well known, non-trivial, and should be considered. In particular, the <u>Bray cycle</u> (May 2016g) is important as we exit the Little Ice Age, the shorter Atlantic Multidecadal Oscillation, and other ocean cycles are important as well (May 2016h).

Besides, modeling precipitation using such a simple natural model and then assigning all the excess precipitation to humans is not reasonable.

Geert Jan van Oldenborgh and colleagues <u>in</u> *Environmental Research Letters*, December 2017, also compare historical precipitation data to the Clausius-Clapyeron (CC) expected rainfall for the current observed warming (Oldenborgh, et al. 2017). This paper is very similar to Emanuel, 2017, and they also conclude that the positive trend in excess precipitation (above the expected CC precipitation) is caused by man-made climate change. They claim that humans made the precipitation 8 to 16 percent more intense. This is like the conclusions of Emanuel, 2017 because they used very similar data and procedures. The conclusion is model-based, and the paper presents no evidence of human influence on the storm. Again, like Emanuel, 2017, the storm is one point, they fit a model to the single point, limit the inputs to CO_2 and their interpretation of "nature," and then attributed everything above their "cyber-nature" to human influence.

We should point out that the Oldenborgh, et al. and Emanuel studies do not agree very well, and the difference is a factor of two. The authors mention this but do not provide an explanation, other than to say the two models must have systemic differences.

They conclude that Houston's flood protection system must be improved and that the frequency of heavy rainstorms in the Houston area has increased. Both conclusions are well supported with data. The problem with both studies is they use a very over-simplified model of nature, then simply assign all excess to humans without sufficient justification.

Historical hurricane data

Our subject is hurricanes and not models of hurricanes. The historical data for Atlantic hurricanes impacting the U.S. suggests that their intensity, as defined by their central pressure, is declining, as shown in Figure 62.

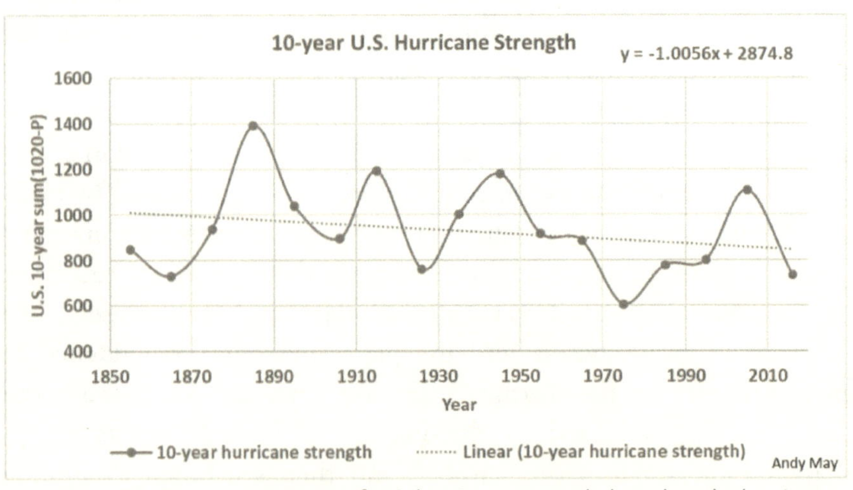

Figure 62. Ten-year averages of U.S. hurricane strength, based on the hurricane central pressure. Data source: NOAA Hurricane Research Division.

In Figure 62, we have subtracted the NOAA estimated hurricane central pressures from 1020 mb and summed the values for all U.S. landfall hurricanes over 10-year periods. These values are plotted at the midpoint of each 10-year period. Each 10-year period starts with the zero year and ends with the ninth year and is plotted at "5." The central hurricane pressure is a good indicator of hurricane strength and is a value available for nearly all hurricanes. We had to estimate the pressure from wind speed for three early hurricanes (in 1918, 1920 and 1953).

By subtracting the value from 1020, we make higher values a strong hurricane and smaller values a weaker hurricane. The trend shows that hurricane strength waxes and wanes over time in a 40- to 60-year pseudo-cycle. Although Irma and Harvey were strong hurricanes, we still seem to be in a hurricane lull in 2017. 2017 hurricanes, from Wikipedia, are included in Figure 62. Table 7 shows the breakdown for the 294 U.S. hurricanes plotted in Figure 62. "P" is the hurricane central pressure from NOAA.

U.S. Hurricanes			
Category in U.S.	Number	sum(1020-P)	Average sum(1020-P)
1	119	4740	40
2	76	3851	51
3	76	5156	68
4	19	1591	84
5	4	452	113
Totals & average strength	294	15790	54

Table 7. U.S. Hurricanes 1850-2017 by category. Data source: NOAA Hurricane Research Division

Pielke Jr. and Lomborg have made similar plots. Pielke's plot for global cyclone and hurricane landfalls is presented as Figure 63. It also shows a decline.

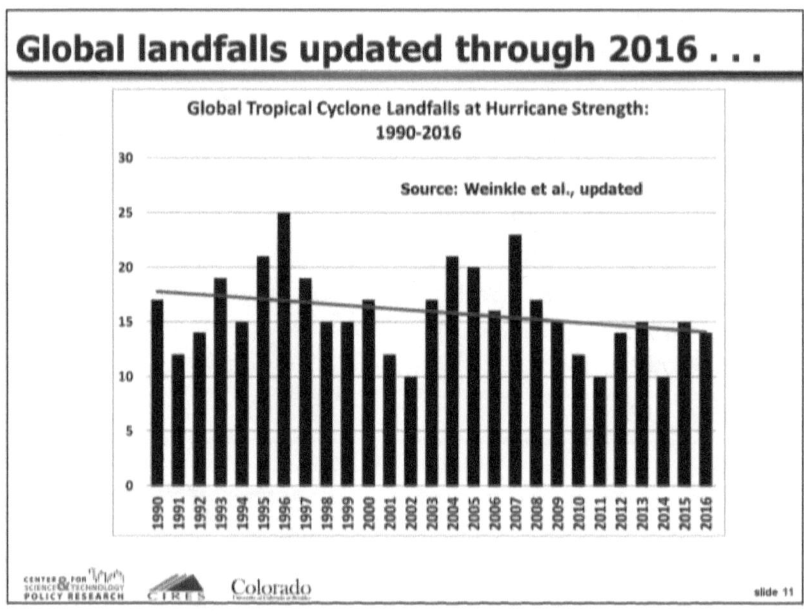

Figure 63. Global tropical cyclone landfalls through 2016. Source: (R. Pielke Jr. 2017), U.S. House of Representatives testimony, used with permission.

Further, Pielke Jr., in Figure 64, also plotted weather-related disaster costs for the world's global economy since 1990. The costs are declining as a percent of GDP.

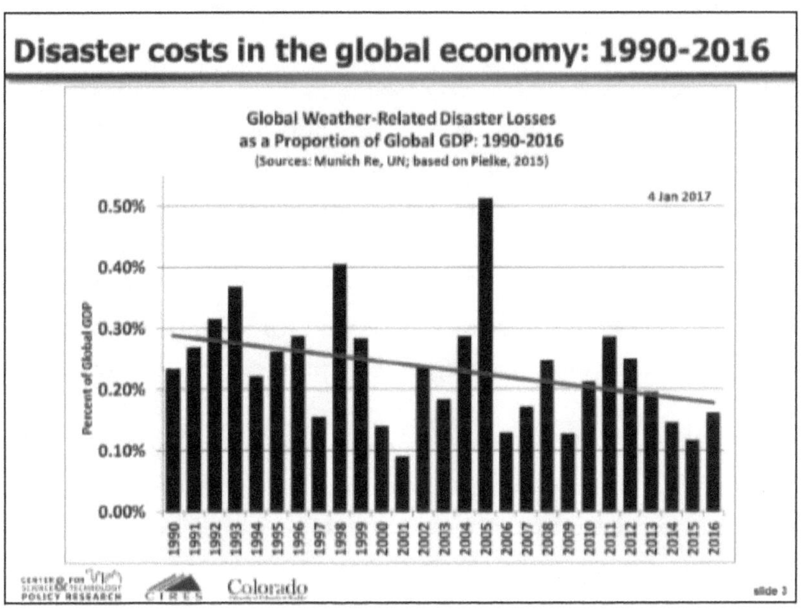

Figure 64. Global weather-related disaster losses as a percent of GDP. Source: (R. Pielke Jr. 2017). House of Representatives testimony, used with permission.

As the world becomes more affluent, GDP goes up and we are better able to handle natural disasters. The data suggests that the severity of hurricanes has gone down recently, not up. It also shows that weather-related disasters are becoming less expensive, as a percentage of GDP, for society to deal with. But, we still hear that they are getting worse:

"Stronger and more frequent hurricanes have become one of the standard exhibits of the global-warming concerns. The Natural Resources Defense Council tells us that 'global warming doesn't create hurricanes, but it does make them

185

stronger and more dangerous.' The group Friends of the Earth proclaims, 'Hurricanes in Florida. Storms in the UK. Extreme weather events are predicted to become more frequent because of climate change.' Greenpeace tells us 'there is strong evidence that extreme weather events—such as hurricanes—are increasing (and becoming more severe and frequent) because of climate change.' The solution offered is invariably CO_2 cuts and adoption of Kyoto." Lomborg, Bjorn. *Cool It* (Kindle Locations 1126-1132).

In 2006, tropical cyclone researchers and forecasters met at a <u>WMO conference</u> (Landsea 2007) and agreed, in part, on the following:

> "1. Though there is evidence both for and against the existence of a detectable anthropogenic [human-caused] signal in the tropical cyclone climate record to date, no firm conclusion can be made on this point."

> "2. No individual tropical cyclone can be directly attributed to climate change."

> "3. The recent increase in societal impact from tropical cyclones has largely been caused by rising concentrations of population and infrastructure in coastal regions."

The theoretical debate over whether hurricanes are increasing in severity is unlikely to be resolved soon. Most alarmist observers end up pointing out how damages from hurricanes are rising dramatically and quickly. This is correct but attributing them to human-caused global warming is wrong. The global costs of weather-related disasters have increased over the past century because the population in dangerous areas has increased, as well as the value of buildings and infrastructure in those areas.

The <u>Clausius-Clapeyron</u> relationship

The recent studies of Hurricane Harvey, cited above, all compare the expected rainfall from the Clausius-Clapeyron relationship to actual data. The Clausius-Clapeyron relationship shows that the equilibrium vapor pressure of water is only related to temperature. As temperature goes up, the equilibrium vapor pressure increases. As the equilibrium vapor pressure just above the water surface increases, more evaporation occurs. So, all else equal (which it never is), as sea surface temperatures go up one-degree Celsius, evaporation increases and the specific (or absolute) humidity in the air goes up about 6 percent. The next assumption is that as the specific humidity goes up precipitation increases by the same amount. These are all reasonable assumptions depending upon air and water circulation patterns and speed, which are unknown.

But, being reasonable is not enough. It is important to compare model results to observations, and none of the studies do that. Risser and Wehner do the best job of tying their results to data, but by assuming that ENSO is the only natural influence on precipitation and temperature; and that man-made CO_2 is the only other influence, their conclusions are suspect. Other factors may be important as well.

Real-world data do not support the Clausius-Clapeyron relationship between specific humidity and temperature. The relationship is sound in the laboratory, but observations show the real-world relationship between humidity, temperature and precipitation is more complex. (Benestad 2017) has reported that the European Centre for Medium-range Weather forecasts (ECMWF) <u>interim re-analysis</u> shows the total volume of water vapor in the atmosphere decreasing by -0.018 kg/m^2

per decade from 1979-2011, and the mean volume is 29 kg/m^2 (see Benestad's Figure 1).

The total water volume is confirmed by RSS (Smith, et al. 2017), who found the atmosphere contains an average of about 28.5 mm of water vapor or 28.5 kg/m^2 of water vapor [the conversion is 1:1 according to (Hanssen, Feut and Kless 2001)], so the decrease seen by ECWMF is about .06 percent/decade or 0.006 percent/year (1979-2011). This is a small decrease, and when error is considered, it is indistinguishable from zero. (Miskolczi 2014) reports that the <u>NOAA R1</u> (NOAA 2017c) dataset shows that global surface air temperature has increased 0.687K between 1948 and 2008, but the water content has decreased by 0.636 percent or -0.0106 percent/year, similar to what is seen in the ECWMF dataset.

The reason why the global specific humidity is decreasing, or perhaps staying the same, as global temperatures are rising is unknown and hotly debated, but it seems to be happening. This observation casts doubt on any study that assumes the Clausius-Clapeyron relationship works alone to control humidity and precipitation. The physical relationship exists, and it will affect humidity to some degree, but there must be other confounding processes at work in the atmosphere.

Global warming will increase flooding

Warming may well result in increasing rain, but it doesn't seem to translate into increasing flooding. A global sample (Kundzewicz, et al. 2009) of 200 rivers showed 27 had increasing high flows, 31 had decreasing flows and 137 showed no change.

In recent times, summer flooding shows no increasing or decreasing trend. Winter flooding has decreased since the Little Ice Age. During the Little Ice Age, it was much colder and ice dams formed in major rivers. This caused catastrophic flooding when temperatures increased, and the dams melted enough to give way. All but two major floods affecting Florence since 1177 occurred before 1844 (B. Lomborg 2007).

Economic losses due to floods have increased, but this is due to more building in flood prone areas. Oddly, it is also due to the increased construction of levees. Levees reduce the size of the floodplain and the extra water storage they provide. People also tend to build on the dry side of levees, a dangerous area.

Yet, modern flood control measures are working overall:

"In 1929, $200 of each $1 million worth of goods got damaged, whereas today only $70 of each $1 million is lost. This indicates that as society has more tangible wealth, while more goods will get damaged in floods, the damage will constitute a smaller and smaller proportion of the total wealth. Overall, floods are not getting more damaging but less." Lomborg, Bjorn. *Cool It* (Kindle Locations 1300-1303).

Lomborg and Pielke also discuss flood prevention costs versus benefits:

"Using the United Kingdom example, for about 0.01 percent of GDP you get a benefit from damage reduction of 0.12 percent of GDP—a benefit-to-cost ratio of 11. From Kyoto, at the cost of 0.5 percent of GDP, you get a benefit of 0.00009 percent of GDP. Or, expressing it in equal terms, a dollar spent on flood management will reduce flooding 1,300 times better than a dollar spent on Kyoto. Flooding is not getting out of hand; the costs are declining compared to total wealth. It is not predominantly a signal of global warming or of increasing heavy rains." Lomborg, Bjorn. *Cool It* (Kindle Locations 1316-1321).

From *The Climate Fix* by Roger Pielke Jr.:

"Research on floods does not support the hypothesis that greenhouse gas emissions have led to a discernible increase in flood losses. Hans Jochim Schellenhuber, a prominent climate scientist and scientific adviser to German chancellor Angela Merkel, coauthored a 2004 paper that summarized the 2001 IPCC assessment and subsequent scientific discussions of floods. It concluded: 'There has been no conclusive and general proof as to how climate change affects flood behavior, in the light of data observed so far.... It is difficult to disentangle the climatic component from the strong natural variability and direct, man-made, environmental changes.'" Pielke Jr., Roger. *The Climate Fix: What Scientists and Politicians Won't Tell You About Global Warming* (p. 174).

Global warming will increase the frequency of droughts and their severity

The Clausius-Clapeyron relationship may not control the amount of precipitation or the global atmospheric specific humidity, but generally we would expect more evaporation with warmer temperatures and, as a result, more precipitation. Specific humidity is determined by the difference between evaporation and precipitation, which are separately controlled by complex processes, and the Clausius-Clapeyron relationship is only one of the processes. Climate models predict 5 percent more precipitation by 2100, which is our only estimate today.

Al Gore, in *An Inconvenient Truth*, and others have told us that drought and hunger in the Sahel region of Africa are caused by global warming (Gore 2007), yet <u>Reuters reported</u> in 2015 (Doyle 2015):

"Rising greenhouse gases have boosted rainfall in the Sahel region of Africa, easing droughts that killed 100,000 people in the 1970s and 1980s, in a rare positive effect of climate change …. The report adds to debate about the causes of a greening of the Sahel region, south of the Sahara Desert from Senegal to Sudan. It said a continued rise in greenhouse gas emissions was likely to help more rainfall in the region in future. … Amounts of rainfall have recovered substantially," said Rowan Sutton, a professor at the National Centre for Atmospheric Science at Britain's Reading University and co-author of the <u>study in the journal Nature Climate Change</u> [(Dong and Sutton 2015)]. And it was a surprise that the increase in greenhouse gases appears to have been the dominant factor, …. Sahel summer rainfall was 0.3 mm (0.01 inch) a day higher from 1996-2011 than the drought period of 1964-93." <u>Reuters, 2015</u>

It turns out that while the Sahel is recovering today, largely because of increases of atmospheric carbon dioxide, the Sahel has had a complex recent history. The Sahel drought from 1968-1973, falls into the normal range of six droughts since 1400AD (Butzer 1983). The drought was a terrible event, but not unusual. Contrary to Gore's claims, additional carbon dioxide in the air and global warming helped the Sahel region and did not harm it.

Chapter 7 Conclusions

Damages from extreme weather events, whether heat waves, cold, droughts, tornados or forest fires are largely a function of the "societal" part of the equation. That is, who lives in the area and how affluent they are. The damages are not due to human-caused climate change. According to Pielke, Jr.:

"Other sorts of extreme events— such as heat waves, cold spells, droughts, and forest fires— have similarly complex stories. In all cases, the societal part of the equation that leads to losses is the dominant factor, not climate change. Yet for some phenomena there are indications that a signal of climate change, and possibly related to greenhouse gas emissions, can be seen in the record of impacts. For instance, some recent research is suggestive that regional warming in the western United States can be associated with increasing forest fires, even in the context of complex patterns of forest management over the past century. If that pattern of warming can be directly attributed to greenhouse gas emissions, then it would be possible to attribute human-caused climate change with a demonstrated impact." Pielke Jr., Roger. *The Climate Fix: What Scientists and Politicians Won't Tell You About Global Warming* (p. 175).

Is the warming of the southwestern part of the U.S., which is making forest fires worse, a result of CO_2 emissions? Or is it a natural climate trend? Even when warming can be identified as a cause somewhere, how do you separate the man-made component from the natural component? This is particularly difficult when we are in the midst of recovering from the Little Ice Age, which is the coldest period in the Holocene, and we are still on a <u>Bray cycle</u> (May 2016g) upswing.

Extreme weather risks are always present, wherever one lives. They vary with location: flooding is a serious risk in the Houston area, drought and wild fires are a serious risk in California, storm surges are a risk in the North Sea. Each location must assess their risks and build appropriate infrastructure to counter the risk as best they can. Trying to prevent so-called "climate change" with a world-wide imagined man-made remedy is a fool's errand. Deal with the specific problem in your own neighborhood and let others deal with theirs. There is no universal solution to the "boogyman." Oops, I mean extreme weather events.

Chapter 7 Summary

There is no evidence that man-made climate change or global warming has increased the probability of extreme weather events. Damages from extreme events have increased in recent years, but this is due to larger populations building expensive buildings and houses in dangerous areas. The strength of storms has decreased, and this is especially true of hurricanes. Once the effect of storms has been corrected for inflation and population increases in sensitive areas, we see a decline in the impact to society.

It is much cheaper and more beneficial to protect sensitive areas from storm hazards than to try and change the weather globally. The impact of local changes to the infrastructure (building seawalls, levees, encouraging people not to build in dangerous locations, etc.) can cost many thousands of dollars less, per dollar saved, than the cost of curtailing CO_2 emissions. This is true even if we accept the idea that curtailing CO_2 will have any measurable effect at all.

If extreme weather is not increasing, what about sea-level rise? Sea level has risen throughout the Holocene, although at a lower rate over the past few thousand years. Is it now accelerating? We examine this idea in the next chapter.

CHAPTER 8

Sea-Level Rise and Glaciers

The <u>IPCC AR5 report</u> (IPCC 2014) has the following to say about the risks of sea-level rise:

"Risks increase disproportionately as temperature increases between 1°–2°C additional warming and become high above 3°C, due to the potential for a large and irreversible sea-level rise from ice sheet loss. For sustained warming greater than some threshold [Current estimates indicate that this threshold is greater than about 1°C (low confidence) but less than about 4°C (medium confidence) sustained global mean warming above preindustrial levels.], near-complete loss of the Greenland ice sheet would occur over a millennium or more, contributing up to 7 m of global mean sea-level rise." AR5, WG2, page 61 (IPCC 2014).

OK, if temperatures increase enough, we could go to a rate of sea-level rise of as much as 7 mm/year (possibly less). We have no idea how much of a temperature increase it would take, but with low to medium confidence it is between 1°C and 4°C.

"Due to sea-level rise projected throughout the 21st century and beyond, coastal systems and low-lying areas will increasingly experience adverse impacts such as submergence, coastal flooding, and coastal erosion (very high confidence). The population and assets projected to be exposed to coastal risks, as well as human pressures on coastal ecosystems, will increase significantly in the coming decades due to population growth, economic development and urbanization (high confidence). The relative costs of coastal adaptation vary strongly among and within regions and countries for the 21st century. Some low-lying developing countries and small island states are expected to face very high impacts that, in some cases, could have associated damage and adaptation costs of several percentage points of GDP." AR5, WG2, page 68 (IPCC 2014b)

A very reasonable statement: as more people build and live on the coast, they are more vulnerable to sea-level rise. Costs to protect these developments vary a lot depending upon where they are. Poorer countries are more at risk.

<u>Kip Hansen reports</u> (K. Hansen, Sea Level: Rise and Fall - Part 1 2017) that The *New York Times* rather breathlessly tells us in 2017:

"A rapid disintegration of Antarctica might, in the worst case, cause the sea to rise so fast that tens of millions of coastal refugees would have to flee inland, potentially straining societies to the breaking point. Climate scientists used to regard that scenario as fit only for Hollywood disaster scripts. But these days, they cannot rule it out with any great confidence. The risk is clear: Antarctica's collapse has the potential to inundate coastal cities across the globe. ... If that ice sheet were to disintegrate, it could raise the level of the sea by more than 160 feet — a potential apocalypse, depending on exactly how fast it happened." — The *NY Times*, <u>Looming Floods, Threatened Cities,</u> a three-part series by Justin Gillis (Gillis 2017)

So, sustained warming over some unknown threshold, perhaps between 1° to 4°C, will cause the Greenland ice sheet to melt in over 1,000 years. Antarctica has ten times as much ice as Greenland, and it will rapidly disintegrate? This paragraph, from the once great *New York Times*, is laughably speculative and dishonest.

This is particularly true because <u>NASA</u> (Vinas 2015) has recently shown that Antarctica is getting colder and gaining ice. This is based on studies by Jay Zwally and colleagues in <u>2015</u> (Zwally, et al. 2015) and <u>2011</u> (Zwally and Giovinetto 2011). Further, Antarctic sea ice extent set records in 2012 and 2014, as discussed by NASA (Ramsayer 2014). Finally, the <u>record cold temperature</u> in Antarctica of -135.8°F (-93.2°C) was set in 2010 and nearly the same temperature was reached in 2013 (Borenstein 2013).

Notice I name my sources and link to peer-reviewed articles, as opposed to the *New York Times* article, which sites anonymous "climate scientists" and "recent computer forecasts." They do allude to Columbia University's Dr. Nicholas Frearson in the previous paragraph, but do not attribute the idea to him. They note the computer forecasts are described as "crude" and "rough" by Robert M. DeConto, University of Massachusetts at Amherst.

The one paper they do cite is (DeConto and Pollard 2016), who use a computer model and the <u>RCP8.5</u> emissions scenario to attempt to show it is possible for Antarctica to contribute 77 cm of sea-level rise by 2100 (9 mm per year) and 12 meters by 2500 (25 mm/year). These are not alarming rates, but unlikely in any case. Dr. Roger Pielke Jr. and (Ritchie and Dowlatabadi 2017) have called the RCP8.5 scenario <u>implausible</u>. All-in-all a very shoddy piece of journalism, but the journalist (Justin Gillis) got a free trip to Antarctica out of it.

The current rate of sea-level rise

The IPCC reports in AR5 WG1 (page 1139):

> "Proxy and instrumental sea-level data indicate a transition in the late 19th century to the early 20th century from relatively low mean rates of rise over the previous two millennia to higher rates of rise (high confidence). It is likely that the rate of global mean sea-level rise has continued to increase since the early 20th century, with estimates that range from 0.000[–0.002 to 0.002] mm yr^{-2} to 0.013 [0.007 to 0.019] mm yr^{-2}. It is very likely that the global mean rate was 1.7 [1.5 to 1.9] mm yr^{-1} between 1901 and 2010 for a total sea-level rise of 0.19 [0.17 to 0.21] m. Between 1993 and 2010, the rate was very likely higher at 3.2 [2.8 to 3.6] mm yr^{-1}; similarly, high rates likely occurred between 1920 and 1950." (IPCC 2013)

Three estimates of sea-level change from the IPCC AR5 WG1 (page 1147) are shown in Figure 65. Figure 65 shows estimates of "eustatic" sea-level rise. This is the rise in sea-level due to an increase in the volume of water in the oceans or the shape of the ocean basins. Today water is added to the oceans by melting continental glaciers or the volume of the water increases as it expands due to raising its temperature.

Isostatic sea-level change is a change in the level of the land, due to geological forces. Sea level can appear to change, but the change is actually the land moving. This is a local thing and not global, like eustatic change. It can be hard to tell if a local change in sea level is eustatic or isostatic.

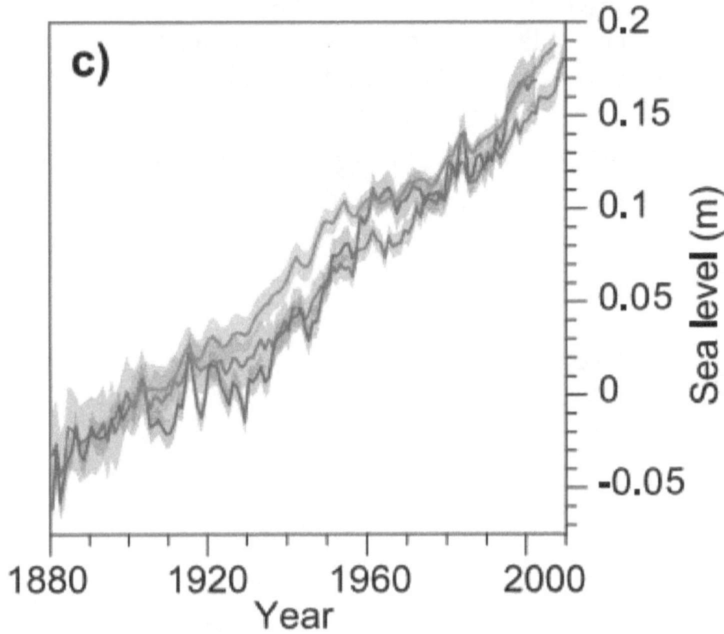

Figure 65. Sea-level change by (Church and White 2011) in orange, (Jerejeva, Grinsted and Moore 2009) in blue, and (Ray and Douglas 2011) in green. (IPCC 2013) used with permission.

All three IPCC estimates of eustatic sea-level rise from 1880 show a steep rise from about 1930 to 1960, followed by a slowing or decline in sea-level rise from 1960 to 1967 and then a steep rise to 1983, another pause for a few years and a rise from 1985 to 2010. Figure 66 shows the components, with the AMO (Atlantic Multidecadal Oscillation) index overlain in green. The AMO is a normalized index of North Atlantic sea-surface temperatures. When it is positive, the North

Atlantic is warm, and when it is negative the North Atlantic is cool. The longer term <u>Bray cycle</u> (May 2016g) is in a warming phase as we come out of the Little Ice Age. This provides a background trend of ocean warming and thus, sea-level rise due to thermal expansion of the ocean water. The sea-level rise component due to thermal expansion is currently about 1 mm/year (0.8 to 1.4) according to the IPCC WG1 AR5 report, page 1151 (IPCC 2013). The slopes of the equations shown in Figure 66 are the rate of increase in sea level for that segment in mm/year. The slope is the coefficient of "x."

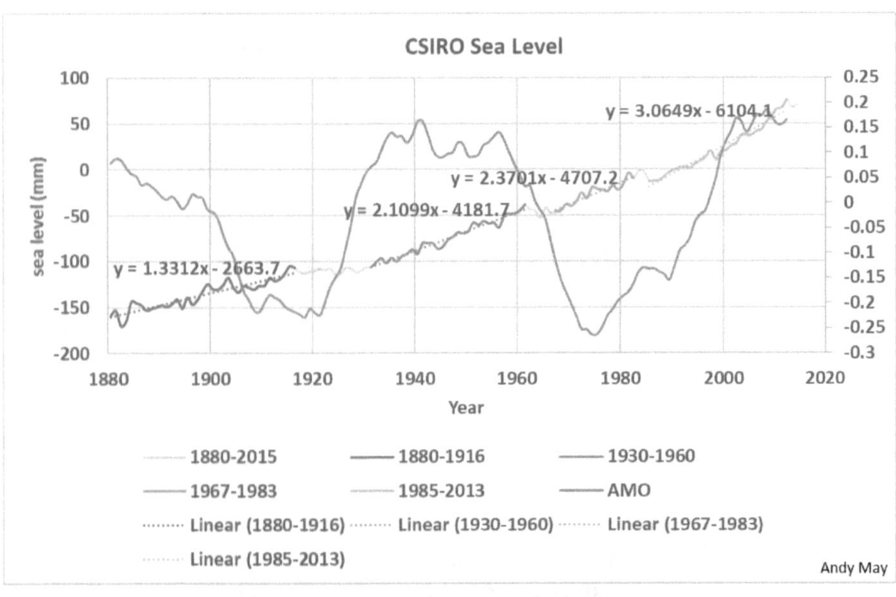

Figure 66. The Commonwealth Scientific and Industrial Research Organization (CSIRO) sea level from 1880 to 2015 and the AMO (Atlantic Multidecadal Oscillation) index. Data sources CSIRO and NOAA.

In Figure 66 the AMO index is shown as is, but perhaps should be lagged a few years. The periods of more rapid eustatic sea-level rise occur when the index is increasing (more thermal energy being taken up by the ocean) or very high. Periods of less rapid sea-level increase are associated with a low or decreasing AMO index (thermal energy being expelled by the ocean).

In the CSIRO record, there is an overall increase in the rate of sea-level rise, from about 2 mm/year (1930-1960) to 3 mm/year (1985-2013). This is a small change and well within the margin of error; the standard deviation of the estimated error (global mean sea-level uncertainty) in the Church and White sea-level data from 1930 to 2013 is 1.9 mm/year. The acceleration, if real, could be an increasing rate of recovery from the Bray cycle low in the Little Ice Age, or due to human greenhouse emissions, or some other ocean process. There is no data that can tell us which it is.

The possible acceleration is modest, and we only have decent data for two AMO lows and recoveries. It is very hard to draw any conclusions with only two values. In another 20 years we will have completed a second AMO high and will know more. The AMO is important, but it is only one of many long-term ocean climate cycles. To read more I recommend Marcia Wyatt's web site (Wyatt 2017).

Other recent estimates of sea-level rise are lower. For example, (Frederikse 2018) estimates that the average rate of sea-level rise from 1958-2014 is 1.5 mm/year ±0.2 mm/year. Another example is (Holgate 2007), where he reports that eustatic sea-level rise is 1.45 ±0.34 mm/year from 1954 to 2003, which is lower than the rate from earlier in the century from 1904-1953 of 2.03 ±0.35 mm/year. Most estimates of eustatic sea-level change are small, and when the accuracy of the estimates is considered, determining any acceleration or deceleration is rampant speculation.

Table 8 shows the components of recent eustatic sea-level rise. The values are from the IPCC WG1 AR5 report (IPCC 2013), NSIDC (National Snow and Ice Data Center 2017) and NASA (Vinas 2015).

Earth's Ice Sheets (1993-2010)							
https://nsidc.org/cryosphere/quickfacts/icesheets.html, NASA, and IPCC WG1 AR5							
	Area	Volume			Current ice volume losses (- = gain)			
					NASA/NSIDC	IPCC AR5	Resulting Range of	
	million sq km	Million cubic km	%	Time to melt at +1° to +4°	Gigatons/yr	Gigatons/yr	Rate of sea level rise mm/yr	Trend
Greenland	1.7	2.9	8.8%	1,000 + years	2.2	157 to 274	0.32 to 0.54	losing ice
Antarctica	14.0	30.0	90.7%	10,000 + years	(-82) to 57	72 to 221	-0.25 to 0.38	stable or increasing ice
Other land sources	0.7	0.2	0.5%	1,000 + years	NA	226 to 301	0.65 to 1.62	losing ice
Thermal Expansion	NA	NA	NA	NA	NA	NA	0.8 to 1.4	Recent warming is about 0.003°C
Total	16.4	33.1					1.52 to 3.9	
Observed (IPCC WG1 AR5)							2.8 to 3.6	Andy May

Table 8. Earth's Ice Sheets and the effect of eustatic sea level rise. Sources: NSIDC, NASA, and the IPCC WG1 AR5 Chapter 13 (Table 13.1 and summary)

In Table 8 we can see the contribution of melting glaciers and ice sheets in Greenland, Antarctica and elsewhere. We have provided estimates by NASA/NSIDC and the IPCC. The NASA values are lower. In terms of mm/year of sea-level rise, these sum to a range of 0.72 to 2.5 mm/year. Once thermal expansion due to a slightly warming ocean is added, the total expected rate of eustatic sea-level rise increases to 1.52 to 3.9 mm/year. We can compare this to the IPCC reported observed rate of 2.8 to 3.6 mm/year. The two ranges (expected and observed) are not alarming, but the IPCC rate is much higher than most other recent sources.

We only have a short record of ocean temperature and it is only to a depth of 2,000 meters. The average depth of the oceans is 3,688 meters. However, making a few reasonable assumptions, we can produce Figure 67 from the JAMSTEC ocean temperature grid. It shows the oceans warming at a rate of 0.003°C per year.

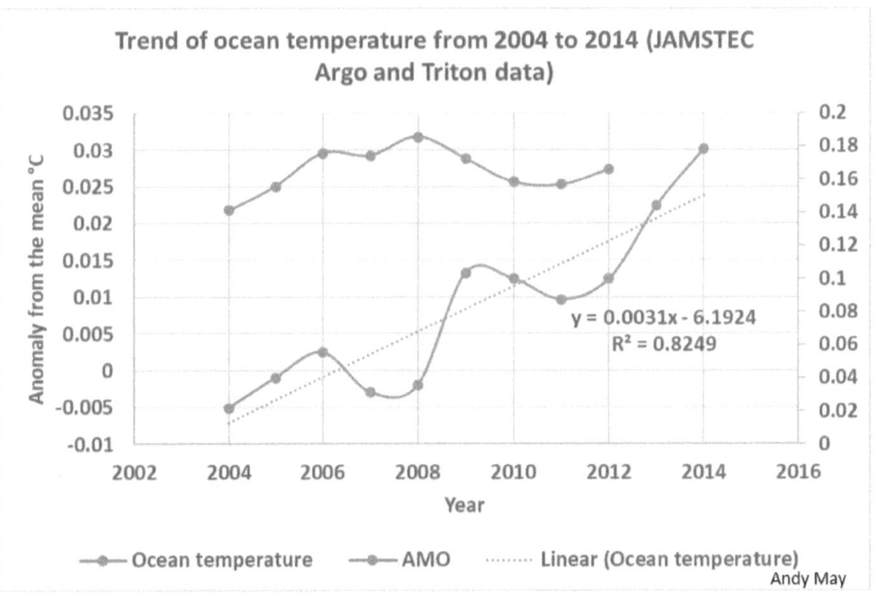

Figure 67. Ocean temperature trend from Argo and TRITON data. Data source JAMSTEC MOAA GPV grid.

The graph shows the ocean temperature change from the surface to 3,688 meters. Zero to 2,000 meters are measured with Argo floats and gridded by JAMSTEC. The ocean temperature at 3,688 meters is assumed to be zero, and an interpolation from 2,000 meters (where the temperature is very close to 2.4°C all the time) is made. Due to the nearly constant global average temperature at 2,000 meters, the interpolation should not affect the trend. The AMO index is overlain on the plot in orange. It explains some of the variability in the whole ocean temperature as we would expect.

(Church and White 2011) write:

> "For 1993–2009 and after correcting for glacial isostatic adjustment, the estimated rate of rise is 3.2 ± 0.4 mm year^{-1} from the satellite data and 2.8 ± 0.8 mm year^{-1} from the in-situ data. The global average sea-level rise from 1880 to 2009 is about 210 mm. The linear trend from 1900 to 2009 is 1.7 ± 0.2 mm year^{-1} and since 1961 is 1.9 ± 0.4 mm year^{-1}."

For the most part, all the better-known (and highest) estimates of recent sea-level rise fall in the range of 3 mm/year +-1 mm/year or so. However, this is a very small number, and the error is large. This is particularly true when we consider we are measuring an average surface affected by waves and tides. The waves are often a meter from trough to peak, and tides can vary several meters or more in 24 hours. Storms greatly confound satellite measurements and fixed tide measurements.

Professor Nils-Axel Mörner has pointed out that that there is a considerable amount of evidence that the rate of sea-level rise is much smaller than reported by the IPCC. He finds evidence from tidal gauges, vegetation and satellite data that sea level has barely risen at all in the last 25 years. In his publication _Sea-Level is not Rising_ (Morner 2012), he lays out the evidence. Since the satellite altimeter data do not agree with the worldwide tidal gauges, our measurements of millimeters of change in sea-level rate are buried in uncertainty.

Kip Hansen, in a series of very well-documented and well-written posts, has explained the complexities involved in measuring sea level and its rise and fall. In his most recent post (part 3 of 3) (K. Hansen 2017b) he summaries his conclusions as follows (I have paraphrased them):

1. Sea-level rise is a threat to coastal cities and very low elevation populated areas (part 1) (K. Hansen 2017).

2. Sea level is not a threat to anything else (part 1).

3. Because land also rises and falls depending upon where you are, local tidal gauges are the most important source of information for communities on the coast (part 2) (K. Hansen 2017c).

4. Local changes in sea level, due to tectonics and tides, are much larger than changes in global sea level and much faster occurring. Eustatic sea-level change is not irrelevant, but it is small and very slow moving (part 2).

5. The tools we use to measure changes in global sea level (satellites and "corrected" tidal gauge records) are only accurate to several centimeters, in practice, and we are trying to use them to measure the change in a dynamic ocean surface. The surface change over an entire year is less than 3 mm, about a tenth of the accuracy of the instruments (part 3). As Hansen points out, we cannot even be 100 percent sure sea level is rising at all.

Land-based sea level measurements are accurate to ± 20 mm and affected by land subsidence or uplift (isostatic changes) as well. The very best and most modern satellites have a measurement accuracy of ± 3 mm (NOAA/NESDIS/STAR 2017) under perfect conditions. They can be affected by weather patterns and problems with orbital decay. Further, sea-level change around the world is not uniform; the global distribution of changes is affected by the ocean cycles mentioned earlier. Beyond these comments, I will encourage the interested reader to read Kip Hansen's posts on how sea-level change is measured and the accuracy of the measurements.

Due to the variation in sea-level rise from place to place, which is mostly due to variations in land movement and tidal ranges over time, local communities should evaluate their own risks, based on local measurements. They need to prepare their community's infrastructure for the specific threats they face. The focus should not be on global sea-level change, this change is small and swamped by measurement errors, but on the local threat to your community. A local investment is far more efficient and thousands of times less expensive than a global mitigation program that may not be effective at all.

Glaciers are retreating

Glaciers have been advancing for most of the past 6,000 years, according to (Mayewski, Rohling, et al. 2004), as the world has cooled from the Holocene Thermal Optimum (May 2015d). Figure 68, from Mayewski's paper, shows some of the evidence. The global cooling from 6,000 years ago is apparent from the worldwide glacial advances plotted in Figure 68c. Present-day is to the left in this plot. In Switzerland (Figure 68d), except for a brief glacial retreat during the Medieval Warm Period 1,000 years ago, glaciers were generally smaller than today before 2,200 years ago.

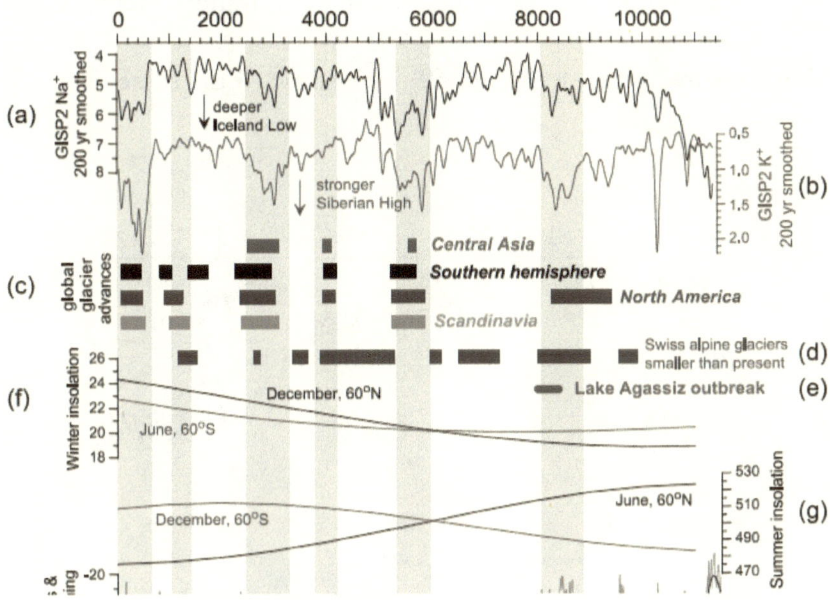

Figure 68. Glacier advance data for the Holocene from (Mayewski, Rohling, et al. 2004). Used with permission.

The dates along the top of Figure 68 are B2K (years before 2000). The green vertical bars in Figure 68 are periods of rapid climate change (RCC). We are currently coming out of the latest RCC, the sixth period of rapid climate change in the Holocene. Figure 68a is a proxy for Icelandic low-pressure events; these events correlate well with northern hemisphere ice sheet growth (Mayewski, Meeker, et al. 1997). Figure 68b is a proxy for the Siberian high-pressure event, which also correlates with ice-sheet growth in the Northern Hemisphere (Mayewski, Meeker, et al. 1997). Figure 68f shows the winter insolation values for the Northern Hemisphere (black) and the Southern Hemisphere (blue). Figure 68g shows the summer insolation for both hemispheres, the summer insolation has decreased in the Holocene in the Northern Hemisphere and increased in the Southern Hemisphere.

Vinós created a figure (Figure 69) with some of the same data.

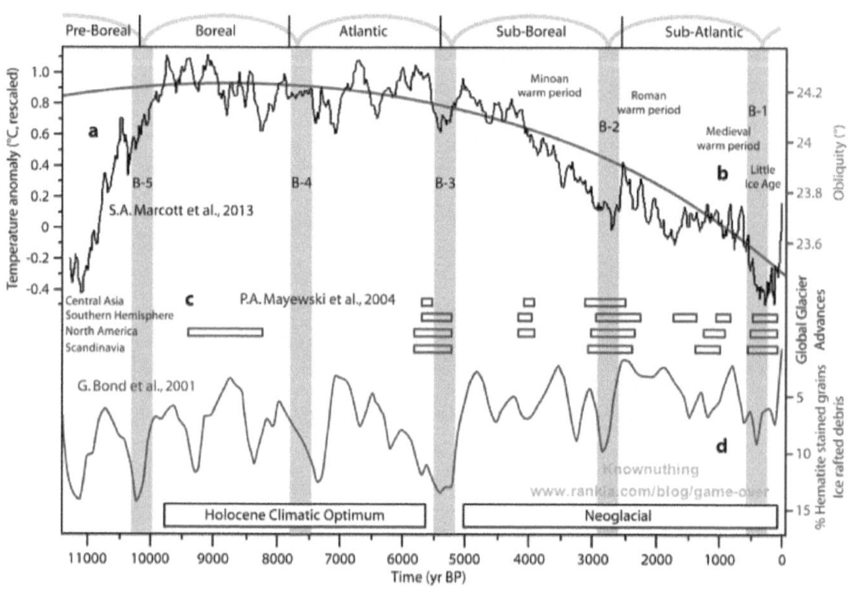

Figure 69. Major climatic events, glacial advances and the Bond hematite-stained grain record for the Holocene. Source (Vinós 2016). Used with permission.

Figure 69 shows the Marcott, et al., 2013 global temperature reconstruction modified to reflect the known temperature difference between the Little Ice Age and the Holocene Thermal Optimum of 1.2°C. This is discussed in Chapter 4 and in the <u>appendix</u> to (Vinós 2016). Chapter 4 goes into some detail on Marcott's reconstruction. The purple curve overlaying the adjusted Marcott reconstruction is the

Earth's obliquity in degrees. It decreases during the Neoglacial period, as discussed in Chapter 4. The Bray (or Hallstatt) cycle cool periods are labeled B-5 through B-1. B-1 is the Little Ice Age. Major global glacial advances are shown with blue boxes. Glaciers are currently retreating as we come out of "B-1" or the Little Ice Age, and this contributes to eustatic sea-level rise.

The red curve in Figure 69 is the abundance of hematite-stained grains found in iceberg debris on the North Atlantic sea floor. The level of this debris increases in colder periods (Bond and Lotti 1995). The scale is inverted to correspond with the temperature curve.

For an alternative global temperature reconstruction, with some problematic proxies removed that shows a 1.2°C difference between the Little Ice Age and the Holocene Thermal Optimum, without adjustment (see Chapter 4, Figure 52). The key point is that the Little Ice Age was the coldest period in the Holocene, and this was largely due to changes in the Earth's orbital tilt, or obliquity.

Lomborg reports in *Cool It*:

"… most glaciers in the Northern Hemisphere were small or absent from nine thousand to six thousand years ago. While glaciers since the last ice age have waxed and waned, they overall seem to have been growing bigger and bigger each time until reaching their absolute maximum at the end of the Little Ice Age. It is estimated that glaciers around 1750 were more widespread on Earth than at any time since the ice ages twelve thousand years ago. So, it is not surprising that as we're leaving the Little Ice Age we are seeing glaciers dwindling. We are comparing them with their absolute maximum over the past ten millennia."

"… with glacial melting, rivers actually increase their water content, especially in the summer, providing more water to many of the poorest people in the world. Glaciers in the Himalayas have been declining significantly since the end of the Little Ice Age and have caused increasing water availability throughout the last centuries, possibly contributing to higher agricultural productivity. But with continuous melting, the glaciers will run dry toward the end of the century. Thus, global warming of glaciers means that a large part of the world can use more water for more than fifty years before they have to invest in extra water storage. These fifty-plus years can give the societies breathing space to tackle many of their more immediate concerns and grow their economies so that they will be better able to afford to build water-storage facilities." Lomborg, Bjorn. *Cool It* (Kindle Locations 884-930).

Global warming will cause excessive sea-level rise

In a previous chapter we discussed the warming of the oceans. We only have significant ocean temperature data since 2004. It is plotted in Figure 67. It shows the temperature in the oceans is rising at about 0.003°C per year currently.

One-third to one-half of the 18 cm rise in sea level that we have seen over the past century (1914-2014, from the CSIRO record plotted in Figure 66) is due to the oceans warming as we come out of the Little Ice Age. Thus, at most, only 12 cm (about 5 inches) are due to melting glaciers and ice sheets.

According to Lomborg in *Cool It:*

"…when water gets warmer, like everything else it expands. Second, runoff from land-based glaciers adds to the ocean water volume. Over the past forty years, glaciers have contributed about 60 percent and water expansion 40 percent of the rise in sea level. In its 2007 report, the UN estimates that sea level will rise about a foot over the rest of the century. While this is not a trivial amount, it is also important to realize that it is certainly not outside historical experience. Since 1860, we have experienced a sea-level rise of about a foot, yet this has clearly not caused major disruptions.

The IPCC cites the total cost for U.S. national protection and property abandonment for more than a three-foot sea-level rise (more than triple what is expected in 2100) at about $5 billion to $6 billion over the century. Considering that the adequate protection costs for Miami would be just a tiny fraction of this cost spread over the century, that the property value for Miami Beach in 2006 was close to $23 billion, and that the Art Deco National Historic District is the second-largest tourist magnet in Florida after Disney World, contributing more than $11 billion annually to the economy, five inches will simply not leave Miami Beach hotels waterlogged and abandoned." Lomborg, Bjorn. *Cool It* (Kindle Locations 956-977).

Sea level rose six inches during the 20th century (see Figure 70) according to the (Church and White 2011). If the IPCC projection for the 21st century is correct, and it rises another 12 to 16 inches, this should not be a problem. We adapted to six inches of sea-level rise with more primitive 20th century technology; another ten inches should not be a problem for 21st century technology. Seawalls, barriers like the Thames Barrier and dikes and levees will be built. Or people may choose to move to higher ground. The key factor is sea level is rising very slowly, and there is plenty of time to adapt.

The IPCC expects the average person in the "standard future" to make $72,700 in the 2080s. If the world decides to mitigate additional CO_2, rather than adapt to additional CO_2, the average person's earnings would decrease to $50,600 according to Lomborg, a reduction of 30 percent. It is possible the environmental world would see more people flooded than the richer, less environmental world because people would be poorer and less able to adapt. In the last century we have lost very little land to higher sea level, simply because the land was valuable enough to protect it with technology.

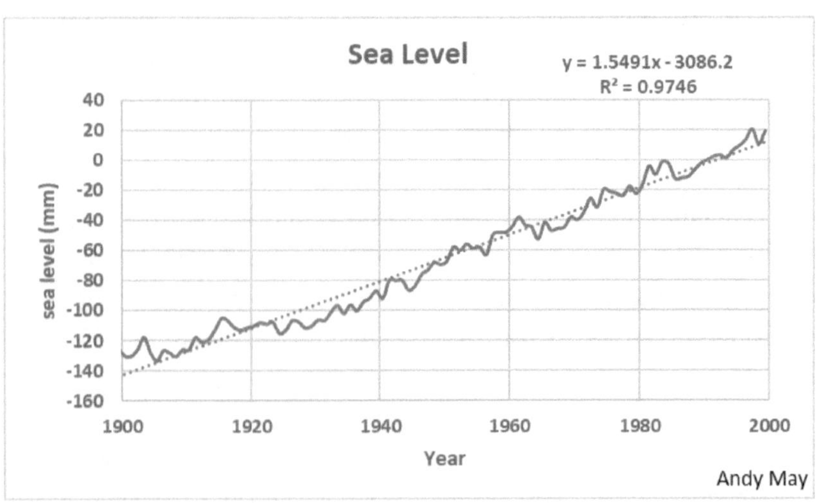

Figure 70. Church and White Sea Level, 20th century.

Figure 71 is a public domain photograph of the famous Thames Barrier that protects London from high tides and North Sea storm surges.

Figure 71. The Thames Barrier, public domain photo.

The largest U.S. death toll, due to a hurricane, was in 1900. The Great Galveston Hurricane was a category 4 storm that made landfall on September 8, 1900, and killed 6,000 to 12,000 people. It is by far the deadliest hurricane in U.S. history. During much of the 19th century, Galveston was the largest city in Texas. By the time of the great storm, however, it was the fourth largest city after Houston, Dallas, and San Antonio. The 15-foot storm surge swept over the entire island and destroyed the city. The survivors mostly moved to Houston and elsewhere in Texas. The island, at the time, only had an eight-foot elevation, and 3,600 homes were washed away.

After the storm the remaining population raised the island's elevation to 17 feet behind a concrete seawall, which was <u>completed in 1911</u>. The seawall has protected Galveston from most subsequent hurricanes, even the monster Hurricane Carla in 1961, which is the strongest hurricane to ever hit the U.S. according to the <u>Hurricane Severity Index</u>.

Figure 72 shows the seawall not long after it was completed.

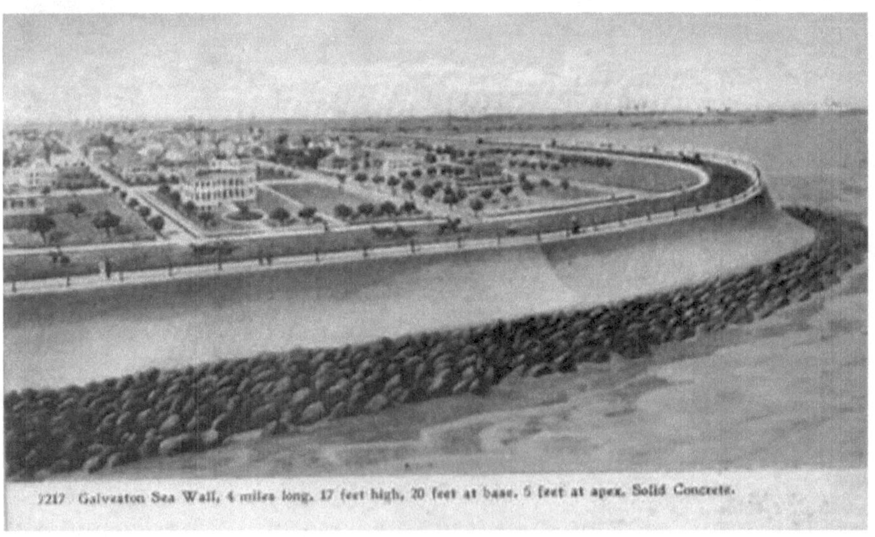

Figure 72. The Galveston seawall in the early 20th century, an undated public domain image.

Figure 73 is a 1905 photo of the wall under construction:

Figure 73. Galveston seawall under construction in 1905, public domain photo, photographer unknown, uploaded to pinterest (link) by Donna Clayton and Ray Gibson..

Figure 74 on the following page shows what the seawall looked like after Hurricane Ike in 2008, the first hurricane to top the wall.

Figure 74. The Galveston seawall in 2008, shortly after Hurricane Ike (photo credit: Aurelia May)

The Army Corps of engineers estimate that $100 million of damage was averted in 1983 from Hurricane Alicia by the seawall. Hurricane Ike (2008) caused a 19-foot storm surge at Galveston, which overtopped the seawall at about 4 pm on September 12, 2008, and caused a great deal of damage. The Bolivar Peninsula, northeast of Galveston, was struck by the strong side of the storm and was submerged under the surge. This destroyed Crystal Beach and surrounding communities and caused flooding of Galveston from the bay side. Since then additional storm surge protections have been proposed, like the so-called "Ike Dike." But, these are still in the planning stages.

Chapter 8 Conclusions

It is true that glaciers are melting today as the world warms after the Little Ice Age. But, the melting glaciers provide much needed water in dry areas. Glaciers reached their maximum Holocene extent during the Little Ice Age, and the glaciers today are still more extensive than they were 6,000 years ago. The melting glaciers, outside of Greenland and Antarctica, contribute 30-50 percent of the expected one-foot rise in sea level over the next 100 years, with thermal expansion of the ocean water contributing most of the remainder. Greenland and Antarctica are not major contributors to sea-level rise, especially if Antarctica is gaining ice, as claimed by Zwally, et al. and NASA.

Worldwide sea-level rise over the next 100 years is not expected to be a problem. Local sea-level rise is a problem for low lying communities, and they should monitor it locally and build local infrastructure to ameliorate its effects. Land can be protected from sea-level rise and severe maritime storms at an affordable cost, or people can move to higher ground.

When one considers that Galveston, Texas, was able to protect itself and rebuild after the devastating 1900 Great Galveston Hurricane in only 11 years, imagine what we can do today, 117 years later. Imagine what we will be able to do 100 years from today. Most analysts suggest that adapting to climate change is better than trying to prevent it. The measures that have been proposed to mitigate climate change, mainly Kyoto and Paris, do very little and are very expensive. Further, we are not even sure that global warming is a problem. Why fix something that may not even matter?

From an economic perspective, the "<u>time value of money</u>" principle tells us it is foolish to invest a serious amount of money today to fix something that may or may not be a problem over 100 years from now. The best investments will be those that benefit us now, and that means adaptation. Mitigation may waste money we will need later to adapt, whether climate change is natural or man-made.

Chapter 8 Summary

Global (or eustatic) sea-level rise is probably happening today, but the rate is very slow and not alarming; perhaps about 2 mm/year to 3 mm/year. The error in measuring eustatic sea-level rise is larger than the rise itself, meaning that we can say it is probably rising, but we cannot be sure, and any rate of rise we determine is not accurate. Further, eustatic sea-level rise is partly expansion and contraction due to natural ocean warming and cooling cycles, thus determining any "acceleration" with the short records we have is problematic.

Sea-level rise is only of concern locally. Local sea-level rise or fall is faster and a function of tidal cycles, eustatic sea-level rise, ocean temperature and local land subsidence or uplift. It can be measured accurately locally and should be. Local solutions, if necessary, such as levees, barriers, dykes and seawalls, are thousands of times more efficient dollar-for-dollar than any attempt at mitigating warming by curtailing CO_2 emissions. Further, adding infrastructure, or moving residents from a dangerous area, does not lower the standard of living globally. Curtailing or eliminating CO_2 emissions does.

CONCLUDING REMARKS

Is a climate catastrophe possible? Sure, they've happened before, and the Younger Dryas cooling and warming event discussed in Chapter 5 (Figure 54) is a good example. Is a man-made climate catastrophe possible? We have seen that the impact of human fossil fuel emissions on our climate has not been measured. The whole idea that a human-caused climate catastrophe might happen in the future is based on climate model predictions. Further, the predictions of the future made by these climate models are not validated, which means they have not matched measurements taken after the predictions were made. Unvalidated model predictions are science fiction.

The sensitivity of our climate to CO_2 or "carbon" emissions is called "ECS" or Equilibrium Climate Sensitivity. It has not been measured, but has been estimated to be 1.5°C to 4.5°C per doubling of CO_2 since the Charney Report was issued in 1979 (Charney, et al. 1979). This large range, a factor of three, has not narrowed since then (Curry 2017).

If anything, the climate science community today is less certain of their estimates of the effect of carbon dioxide emissions than they were in 1979. The warming observed recently has been half of what was projected using previous "most likely" estimates of ECS of about 3°C. Attempts to measure ECS with an observation-based approach are not conclusive, but they are trending lower, as many recent estimates are below 2°C (Curry 2017).

Even though the IPCC AR5 (IPCC 2013) declined to select a "best estimate" of the effect of CO_2 on climate due to the discrepancy between global climate model estimates and observations, the entire

IPCC AR5 WG3 report on "Mitigation of Climate Change" was based upon an ECS of 3.0°C (Edenhofer, et al. 2014) (Curry 2017).

Thus, the IPCC calculations of the "safe" and "dangerous" amounts of CO_2 emissions are based upon science fiction - model results that do not match observations. What we observe today is that warming is slower than predicted, and it appears to be affected by natural forces, such as ocean oscillations and solar variability, more than predicted.

The science of climate change is anything but settled. Before spending our money and prosperity on dreams of controlling the climate, we should at least be sure what we are buying will work. If nothing else, we should wait until the effect of carbon dioxide emissions on the climate can be measured.

ABOUT THE AUTHOR

Andy May is a writer, blogger and author living in The Woodlands, Texas. He enjoys golf and traveling in his spare time. He is also an editor for the popular climate change blog Wattsupwiththat.com, where he has published numerous posts and is the author or co-author of seven peer-reviewed papers on various geological, engineering and petrophysical topics. He has also written about computers and computer software. His personal blog is andymaypetrophysicist[.]com.

He retired from a 42-year career in petrophysics in 2016. Most of his petrophysical work was for several oil and gas companies worldwide. He has worked in exploring, appraising and developing oil and gas fields in the U.S., Argentina, Brazil, Indonesia, Thailand, China, the U.K. North Sea, Canada, Mexico, Venezuela and Russia. He helped discover and appraise several large fields including Block 0436 offshore of China; Gryphon Field in the UK North Sea; Nansen and Boomvang in the Gulf of Mexico; and Natuna D-Alpha in Indonesia.

Late in his career, he worked on unconventional shale oil and gas petrophysics and developed many unique techniques for evaluating these difficult reservoirs. In cooperation with Professor Mike Lovell (University of Leicester in the U.K.) he developed a one-week course in shale reservoir petrophysics. Andy has a B.S. in Geology from the University of Kansas.

WORKS CITED

Allansson, Marie, Erik Melander, and Lotta Themnér. 2017. "Organized violence, 1989-2016." *Journal of Peace Research* 54 (4).

Alley, R. B. 2004. "GISP2 Ice Core Temperature and Accumulation Data." *IGBP PAGES/World Data Center for Paleoclimatology* (NOAA/NGDC Paleoclimatology Program, Boulder CO, USA.   ). ftp://ftp.ncdc.noaa.gov/pub/data/paleo/icecore/greenland/summit/gisp2/isotopes/gisp2_temp_accum_alley2000.txt.

Alley, R. B. 2000. "The Younger Dryas cold interval as viewed from Central Greenland." *Quaternary Science Reviews* 19: 213-226. http://klimarealistene.com/web-content/Bibliografi/Alley2000%20The%20Younger%20Dryas%20cold%20interval%20as%20viewed%20from%20central%20Greenland%20QSR.pdf.

American Museum of Natural History. 2017. "Mass Extinction." https://www.amnh.org/exhibitions/dinosaurs-ancient-fossils-new-discoveries/extinction/mass-extinction/.

Amos, Jonathan. 2017. "Defining a true 'pre-industrial' climate period." *BBC News*, January 25. http://www.bbc.com/news/science-environment-38745937.

Argo. 2017. "What is Argo?" http://www.argo.ucsd.edu/.

Arsenault, Chris. 2015. "Global insecurity and refugee crisis linked to climate change: expert." *Reuters*, August 26.

https://www.reuters.com/article/us-climatechange-security-aid/global-insecurity-and-refugee-crisis-linked-to-climate-change-expert-idUSKCN0QV0ER20150826.

Atwood, Todd C., Elizabeth Peacock, Melissa McKinney, Kate Lillie, Ryan Wilson, David Douglas, Susanne Miller, and Pat Terletzky. 2016. "Rapid Environmental Change Drives Increased Land Use by an Arctic Marine Predator." *PLOS*. http://journals.plos.org/plosone/article?id=10.1371%2Fjournal.pone.0155932.

Barnosky, A. D. 2008. "Megafauna biomass tradeoff as a driver of Quaternary and future extinctions." *PNAS* 105: 11543-11548. http://www.pnas.org/content/pnas/105/Supplement_1/11543.full.pdf.

Baron, Gemma L., Vincent A. A. Jansen, Mark J. F. Brown, and Nigel E. Raine. 2016. "Pesticide reduces bumblebee colony initiation and increases probability of population extinction." *Nature Ecology and Evolution* 1308-1316. https://www.nature.com/articles/s41559-017-0260-1.

BBC. 2014. "Big Five mass extinction events." *Nature prehistoric life*, October. http://www.bbc.co.uk/nature/extinction_events.

Bendle, J., and A. Rosell-Mele. 2007. "High-resolution alkenone sea surface temperature variability on the North Icelandic Shelf: implications for Nordic Seas palaeoclimatic development during the Holocene." *The Holocene* 17 (1): 9-24. http://www.realscience.org.uk/Holocene.pdf.

Benestad, Rasmus E. 2017. "A mental picture of the greenhouse effect." *Theoretical and Applied Climatology* 128: 679-688.

https://link.springer.com/article/10.1007/s00704-016-1732-y.

Bereiter, Bernhard, Sarah Shackleton, Daniel Baggenstos, Kenji Kawamura, and Jeff Severinghaus. 2018. "Mean global ocean temperatures during the last glacial transition." *Nature* 553: 39-44. https://www.nature.com/articles/nature25152.

Berko, J., D. Ingram, S. Saha, and J. Parker. 2014. *Deaths Attributed to Heat, Cold, and other Weather Events in the United States, 2006-2010.* U.S. Department of Health and Human Services. https://www.cdc.gov/nchs/data/nhsr/nhsr076.pdf.

Bernstein, Paul, W. David Montgomery, Bharat Ramkrishnan, and Sugandha D. Tuladhar. 2017. *Impacts of Greenhouse Gas Regulations On the Industrial Sector.* NERA Economic Consulting, Washington D.C.: American Council for Capital Formation. http://cdn.accf.org/wp-content/uploads/2017/03/170316-NERA-ACCF-Full-Report.pdf.

Billings, Lee. 2013. "Fact or Fiction?: We Can Push the Planet into a Runaway Greenhouse Apocalypse." *Scientific American.* https://www.scientificamerican.com/article/fact-or-fiction-runaway-greenhouse/.

Bond, Gerard, and Rusty Lotti. 1995. "Iceberg Discharges into the North Atlantic on Millennial Time Scales During the Last Glaciation." *Science* 267. http://science.sciencemag.org/content/267/5200/1005.

Borenstein, Seth. 2013. "Antarctica Sets Cold Record of -135.8 Degrees." *The Weather Channel*, December 10. https://weather.com/news/news/antarctica-sets-cold-record-20131209.

Bosello, F., R. Roso, and R. Tol. 2005. "Economy-Wide Estimates of the Implications of Climate Change: Human Health." https://www.econstor.eu/bitstream/10419/74106/1/NDL2005-097.pdf.

Botkin, Daniel. 2016. *25 Myths that are Destroying the Environment.* Taylor Trade Publishing. https://www.amazon.com/dp/B01L9XMFM2/ref=dp-kindle-redirect?_encoding=UTF8&btkr=1.

BP. 2017. *BP Energy Outlook - 2017 edition.* BP. https://www.bp.com/en/global/corporate/energy-economics/energy-outlook/energy-outlook-downloads.html.

Brannen, Peter. 2017. "Earth Is Not in the Midst of a Sixth Mass Extinction." *The Atlantic*, June 13. https://www.theatlantic.com/science/archive/2017/06/the-ends-of-the-world/529545/.

Brewer, S., J. Guiot, and D. Barboni. 2014. "Pollen Methods and Studies, Use of Pollen as Climate Proxies." *Earth Systems and Environmental Sciences* 3: 2497-2508. https://www.sciencedirect.com/science/article/pii/B9780444536433001801.

Bromaghin, Jeffrey F., Trent McDonald, Ian Stirling, Andrew Derocher, Evan Richardson, Eric Regehr, David Douglas, George Durner, Todd Atwood, and Steven Amstup. 2015.

"Polar bear population dynamics in the southern Beaufort Sea during a period of sea ice decline." *Ecological Applications* 634-651. http://onlinelibrary.wiley.com/doi/10.1890/14-1129.1/abstract.

Buis, Alan. 2010. "NASA Study Finds Atlantic 'Conveyor Belt' Not Slowing." March 25. https://www.nasa.gov/topics/earth/features/atlantic2010032 5.html.

Bulatao, R. A. 1993. "Mortality by Cause, 1970 to 2015." Edited by Gribble and Preston. *The Epidemiological Transition: Policy and Planning Implications for Developing Countries.* Washington, D.C.: National Academy Press. https://books.google.com/books?hl=en&lr=&id=_EQrAAA AYAAJ&oi=fnd&pg=PA42&dq=Mortality+by+cause,+1970 +to+2015&ots=3bveBK5H6c&sig=6u2h0ypLYWQK2SeYxr UZ6Ja0YHY#v=onepage&q=Mortality%20by%20cause%2C %201970%20to%202015&f=false.

Busby, Joshua, Kerry Cook, Edward Vizy, Todd Smith, and Mesfin Bekalo. 2014. "Identifying hot spots of security vulnerability associated with climate change in Africa." *Climate Change* 124: 717-731. https://www.researchgate.net/profile/Kerry_Cook2/publicati on/264859757_Identifying_hot_spots_of_security_vulnerabili ty_associated_with_climate_change_in_Africa/links/56337e2 d08ae758841120e87.pdf.

Busby, Joshua, Todd Smith, Kaiba White, and Shawn Strange. 2013. "Climate Change and Insecurity: Mapping Vulnerability in

Africa." *International Security* 37 (4). https://www.mitpressjournals.org/doi/abs/10.1162/ISEC_a _00116#authorsTabList.

Butzer, K. 1983. "Paleo-environmental perspectives on the Sahel drought of 1968–73." *GeoJournal* 369-374. https://link.springer.com/article/10.1007%2FBF00241460.

Cama, Timothy, and Devin Henry. 2017. "Trump: We are getting out of Paris climate deal." *The Hill*, June 1. http://thehill.com/policy/energy-environment/335955-trump-pulls-us-out-of-paris-climate-deal.

Carter, R., R. McKitrick, V. Courtillot, I. Plimer, F. Dyson, M. Ridley, C. Essex, et al. 2015. *THE SMALL PRINT What the Royal Society Left Out.* The Global Warming Policy Foundation. https://www.thegwpf.org/content/uploads/2015/03/Shortg uide.pdf.

Cass, Oren. 2016. "Another Obama Legacy: Americans Will Pay Billions for a Useless Climate Agreement." *National Review.* http://www.nationalreview.com/article/434412/paris-climate-agreement-americans-foot-bill-no-effect-climate.

CDC. 2017. *Leading Causes of Death.* Hyattsville, MD: National Center for Health Statistics. https://www.cdc.gov/nchs/fastats/leading-causes-of-death.htm.

Ceballos, Gerardo, Paul R. Ehrlich, Anthony D. Barnosky, Andrés García, and Robert M. Pringle. 2015. "Accelerated modern human–induced species losses: Entering the sixth mass extinction." *Science Advances.* http://advances.sciencemag.org/content/1/5/e1400253.full.

Charney, J., A. Arakawa, D. Baker, B. Bolin, R. Dickinson, R. Goody, C. Leith, H. Stommel, and C. Wunsch. 1979. *Carbon Dioxide and Climate: A Scientific Assessment.* National Research Council, Washington DC: National Academy of Sciences. http://www.ecd.bnl.gov/steve/charney_report1979.pdf.

Chen, Yunguang, and Marc A.C. Hafstead. 2016. *Using a Carbon Tax to Meet US International Climate Pledges.* Resources for the Future, Washington, D.C.: Resources for the Future. http://www.rff.org/files/document/file/RFF-DP-16-48.pdf.

ChinaPower. 2017. "Developing or Developed? Assessing Chinese life expectancy." https://chinapower.csis.org/life-expectancy/.

Chopra, Deepak. 2005. "Is the Human Race a Planetary Cancer?" *Huffpost.* https://www.huffingtonpost.com/deepak-chopra/is-the-human-race-a-plane_b_7092.html.

Church, John A., and Neil J. White. 2011. "Sea-Level Rise from the Late 19th to the Early 21st Century." *Surveys in Geophysics* 32: 585-602. https://link.springer.com/article/10.1007/s10712-011-9119-1.

Cline, Eric H. 2015. *1177 B.C.: The Year Civilization Collapsed (Turning Points in Ancient History).* Princeton Univerisity Press. https://www.amazon.com/1177-B-C-Civilization-Collapsed-Turning-

ebook/dp/B013VPYYGQ/ref=sr_1_1?ie=UTF8&qid=1497
005929&sr=8-1&keywords=1177+BC.

Crockford, S. J. 2017b. "Testing the hypothesis that routine sea ice
coverage of 3-5 mkm2 results in a greater than 30% decline in
population size of polar bears (Ursus maritimus)." *PeerJ*.
https://polarbearscience.files.wordpress.com/2017/01/crock
ford-2017_polar-bear-sea-ice-hypothesis.pdf.

Crockford, Susan. 2017. "IUCN Specialist Group now rejects polar
bear numbers it used for 2015 IUCN Red List review." May 1.
https://polarbearscience.com/2017/05/01/iucn-specialist-
group-now-rejects-polar-bear-numbers-it-used-for-2015-iucn-
red-list-review/.

Curry, J. 2017. *Climate Models for the layman*. GWPF Reports.
https://www.thegwpf.org/content/uploads/2017/02/Curry-
2017.pdf.

Dayaratna, Kevin D., Nicolas D. Loris, and and David W. Kreutzer.
2016. *Consequences of Paris Protocol: Devastating Economic Costs,
Essentially Zero Environmental Benefits*. Heritage Foundation,
Heritage Foundation.
https://www.heritage.org/sites/default/files/2017-
09/BG3080.pdf.

DeConto, R., and D. Pollard. 2016. "Contribution of Antarctica to past
and future sea-level rise." *Nature*.
https://www.geo.umass.edu/climate/papers2/DeConto2016.
pdf.

deMenocal, P. B. 2001. "Cultural Responses to Climate Change During
the Late Holocene." *Science* 292: 667-673.

http://www.jstor.org/stable/3083536?seq=1#page_scan_tab
_contents.

DeMenocal, P., J. Ortiz, T. Guilderson, and M. Sarnthein. 2000.
"Coherent High- and Low-Latitude Climate Variability During
the Holocene Warm Period." *Science* 2198-2202.
https://www.researchgate.net/publication/12453679_Cohere
nt_High-_and_Low-
Latitude_Climate_Variability_During_the_Holocene_Warm_
Period.

Dixon, P., D. Brommer, B. Hedquist, A. Kalkstein, G. Goodrich, J.
Wlter, C. Dickerson, S. Penney, and R. Cerveny. 2005. "HEAT
MORTALITY VERSUS COLD MORTALITY A Study of
Conflicting Databases in the United States." *AMERICAN
METEOROLOGICAL SOCIETY* 937-943.
https://journals.ametsoc.org/doi/pdf/10.1175/BAMS-86-7-
937.

Dong, Buwen, and Rowan Sutton. 2015. "Dominant role of
greenhouse-gas forcing in the recovery of Sahel rainfall." *Nature
Climae Change* 757-760.
https://www.nature.com/articles/nclimate2664.

Dong, Shenfu, J. Sprintall, and Sarah Gille. 2006. "Location of the
Antarctic Polar Front from AMSR-E Satellite Sea Surface
Temperature Measurements." *Journal of Physical Oceanography* 36:
2075-2089.
http://pordlabs.ucsd.edu/sgille/pub_dir/dong_et_al_jpo06.p
df.

Doyle, Alister. 2015. "Climate Change boosts rain in Africa's Sahel region: study." *Reuters*, June 1. https://www.reuters.com/article/us-climatechange-sahel/climate-change-boosts-rain-in-africas-sahel-region-study-idUSKBN0OH2UY20150601.

Drake, Nadia. 2015. "Will Humans Survive the Sixth Great Extinction?" *National Geographic*. https://news.nationalgeographic.com/2015/06/150623-sixth-extinction-kolbert-animals-conservation-science-world/.

Edenhofer, O., R. Pichs-Madruga, Y. Sokona, S. Kadner, J. C. Minx, S. Brunner, S. Agrawala, et al. 2014. "2014: WG3 Technical Summary." In *Climate Change 2014: Mitigation of Climate Change. Contribution of Working Group III to the Fifth Assessment Report of the Intergovernmental Panel on Climate Change*, by O. Edenhofer, R. Pichs-Madruga, Y. Sokona, E. Farahani, S. Kadner, K. Seyboth, A. Adler, et al. Cambridge: Cambridge University Press. https://www.ipcc.ch/pdf/assessment-report/ar5/wg3/ipcc_wg3_ar5_technical-summary.pdf.

Ehrlich, Paul. 1968. *The Population Bomb*. Buccaneer Books. https://www.amazon.com/Population-Bomb-Paul-Ehrlich-1995-12-01/dp/B01JXQJGRO/ref=tmm_hrd_swatch_0?_encoding=UTF8&qid=1516638475&sr=8-1.

EIA. 2017b. *Annual Energy Outlook 2017*. EIA. https://www.eia.gov/outlooks/aeo/index.php.

—. 2012. "Electricity's share of U.S. delivered energy has risen significantly since 1950." *Today in Energy*, March 2. https://www.eia.gov/todayinenergy/detail.php?id=5230.

EIA. 2017. "What is U.S. electricity generation by energy source?" https://www.eia.gov/tools/faqs/faq.php?id=427&t=3.

Emanuel, Kerry. 2017. "Assessing the present and future probability of Hurricane Harvey's rainfall." *PNAS*. http://www.pnas.org/content/early/2017/11/07/171622211 4.short.

Energy Policy Institute at the University of Chicago. 2017. "Air Quality-Life Index (AQLI)." https://aqli.epic.uchicago.edu/wp-content/uploads/2017/09/AQLI_1Pager_Final.pdf.

EPA. 2017b. *Climate Change Indicators: High and Low Temperatures*. EPA. https://www.epa.gov/climate-indicators/climate-change-indicators-high-and-low-temperatures.

—. 2017. "Progress Cleaning the Air and Improving People's Health." https://www.epa.gov/clean-air-act-overview/progress-cleaning-air-and-improving-peoples-health.

Epstein, Alex. 2014. *The Moral Case for Fossil Fuels*. Penguin Publishing Group. https://www.amazon.com/Moral-Case-Fossil-Fuels-ebook/dp/B00INIQVJA/ref=sr_1_1?s=digital-text&ie=UTF8&qid=1516628309&sr=1-1&keywords=The+Moral+Case+for+fossil+fuels.

ExxonMobil. 2017. "2017 Outlook for Energy: A View to 2040." http://corporate.exxonmobil.com/en/energy/energy-outlook/download-the-report/download-the-outlook-for-energy-reports.

FAO Statistics Division. 2017. "FAOSTAT." http://www.fao.org/faostat/en/#home.

FAO. 2017b. "The State of Food Security and Nutrition in the World - 2017." http://www.fao.org/state-of-food-security-nutrition/en/.

FAO. 2017c. "World Food Situation." http://www.fao.org/worldfoodsituation/csdb/en/.

Field, C.B., V.R. Barros, K.J. Mach, M.D. Mastrandrea, M. van Aalst, W.N. Adger, D.J. Arent, et al. 2014. "WG2 Technical summary." In *Climate Change 2014: Impacts, Adaptation, and Vulnerability. Part A: Global and Sectoral Aspects. Contribution of Working Group II to the Fifth Assessment Report of the Intergovernmental Panel on Climate Change*, by C.B. Field, V.R. Barros, D.J. Dokken, K.J. Mach, M.D. Mastrandrea, T.E. Bilir, M. Chatterjee, et al., edited by C.B., V.R. Barros, D.J. Dokken, K.J. Mach, M.D. Mastrandrea, T.E. Bilir, M. Chatterjee, K.L. Ebi, Y.O. Estrada, R.C. Genova, B. Girma, E.S. Kissel, A.N. Levy, S. MacCracken, P.R. Mastrandrea, and L.L. White Field, 35-94. Cambridge: Cambridge University Press. http://www.ipcc.ch/pdf/assessment-report/ar5/wg2/WGIIAR5-TS_FINAL.pdf.

Foster, G. 2013. "The Tick." March 22. https://tamino.wordpress.com/2013/03/22/the-tick/.

Frederikse, Thomas. 2018. "A Consistent Sea-Level Reconstruction and Its Budget on Basin and Global Scales over 1958–2014." *American Meteorological Society Journal.* https://journals.ametsoc.org/doi/abs/10.1175/JCLI-D-17-0502.1.

Fuelfix. 2016. "Ethanol blending shrinking margins for refiners." *Fuelfix*. http://fuelfix.com/blog/2016/05/13/ethanol-blending-shrinking-margins-for-refiners/.

Ghosh, Pallab. 2017. "Hawking says Trump's climate stance could damage Earth." *BBC*. http://www.bbc.com/news/science-environment-40461726.

Gillis, Justin. 2017. "Antarctic Dispatches." *New York Times*, May 18. https://www.nytimes.com/interactive/2017/05/18/climate/antarctica-ice-melt-climate-change.html.

Gittleman, John L. 2017. "Extinction." In *Encyclopaedia Britannica*. Encyclopaedia Britannica Inc. https://www.britannica.com/science/extinction-biology.

Gleditsch, Nils Petter, Peter Wallensteen, Mikael Eriksson, Margareta Sollenberg, and Håvard Strand. 2002. "Armed Conflict 1946-2001: A New Dataset." *Journal of Peace Research* 39 (5).

Gore, Al. 2007. *An Inconvenient Truth: The Crisis of Global Warming*. Viking Books.

Gray, Louis. 2013. "David Attenborough – Humans are plague on Earth." *The Telegraph*. http://www.telegraph.co.uk/news/earth/earthnews/9815862/Humans-are-plague-on-Earth-Attenborough.html.

Hager, Thomas. 2008. *The Alchemy of Air: A Jewish Genius, a Doomed Tycoon, and a Scientific Discovery That Fed the World but Fueled the Rise of Hitler*. Random House LLC.

https://www.amazon.com/Alchemy-Air-Jewish-Scientific-Discovery-ebook/dp/B001EUGCTS/ref=sr_1_1?ie=UTF8&qid=15202 51088&sr=8-1&keywords=haber+bosch.

Hannon, Renee. 2018. "Modern Warming - Climate Variablity or Climate Change." *Wattsupwiththat.com*, March 28. https://wattsupwiththat.com/2018/03/28/__trashed-6__trashed/.

Hansen, James. 2009. *Storms of My Grandchildren: The Truth about the Coming Climate Catastrophe and Our Last Chance to Save Humanity.* Bloomsbury. https://www.amazon.com/Storms-My-Grandchildren-Catastrophe-Humanity-ebook/dp/B002Z8IWLO/ref=sr_1_fkmr0_1?ie=UTF8&qid =1472925900&sr=8-1-fkmr0&keywords=Hansen%2C+J.+2009+Storms+of+my+g randchildren%3A+The+truth+about+the+coming+climate+ catastrophe+and+our+la.

Hansen, Kip. 2017. "Sea Level: Rise and Fall - Part 1." https://wattsupwiththat.com/2017/09/13/sea-level-rise-and-fall-part-1/.

Hansen, Kip. 2017c. "Sea Level: Rise and Fall - Part 2 - Tide Gauges." https://wattsupwiththat.com/2017/10/07/sea-level-rise-and-fall-part-2-tide-gauges/.

Hansen, Kip. 2017b. "Sea Level: Rise and Fall - Part 3 - Computational Hubris." https://wattsupwiththat.com/2017/12/19/sea-level-rise-and-fall-part-3-computational-hubris/.

Hanssen, Ramon, A. Feut, and R. Kless. 2001. "Comparison of Precipitable Water Vapor Observations by Spaceborne Radar Interferometry and Meteosat 6.7-micrometer Radiometry." *J. of Atmospheric and Oceanic Technology.* https://journals.ametsoc.org/doi/pdf/10.1175/1520-0426%282001%29058%3C0756%3ACOPWVO%3E2.0.CO%3B2.

Harbin, Christine. 2017. "The Paris climate agreement was a terrible deal for the US." *Washington Examiner,* June 5. http://www.washingtonexaminer.com/the-paris-climate-agreement-was-a-terrible-deal-for-the-us/article/2624974.

Holgate, S. 2007. "On the decadal rates of sea level change during the twentieth century." *Geophysical Research Letters* 34 (1). http://onlinelibrary.wiley.com/doi/10.1029/2006GL028492/abstract.

Holt, B. M., and V. Formicola. 2008. "Hunters of the Ice Age: The biology of Upper Paleolithic people." *Am J Phys Anthropology* 47: 70-99. https://www.ncbi.nlm.nih.gov/pubmed/19003886.

Hu, Aixue, G. Meehl, W. Han, and J. Yin. 2009. "Transient response of the MOC and climate to potential melting of the Greenland Ice Sheet in the 21st century." *Geophysical Research Letters* 36 (10). http://onlinelibrary.wiley.com/doi/10.1029/2009GL037998/full.

Huguet, Carme, Jung-Hyun Kim, Jaap S. Sinninghe Damste, and Stefan Schouton. 2006. "Reconstruction of sea surface temperature variations in the Arabian Sea over the last 23 kyr using organic proxies (TEX86 and U37K')." *Paleoceanography and Paleoclimatology.* http://onlinelibrary.wiley.com/doi/10.1029/2005PA001215/f ull.

Idso, C. 2017b. "CO2Science.org." http://co2science.org/.

Idso, C. 2017c. "Interaction of CO2 and Water Stress on Plant Growth (Agricultural Species)." *CO2Science.org.* http://co2science.org/subject/g/growthwaterag.php.

Idso, Craig. 2017. "Energy, Agriculture and Rising Atmospheric CO2." *America First Energy.* Houston: Heartland Institute. http://americafirstenergy.org/.

—. 2011. *The Many Benefits of Atmospheric CO2 Enrichment.* Perfect Paperback. https://www.amazon.com/Many-Benefits-Atmospheric-CO2-Enrichment/dp/0981969429/ref=sr_1_1?ie=UTF8&qid=151 1639755&sr=8-1&keywords=The+many+benefits+of+atmospheric+CO2+e nrichment.

—. 2013. *The Positive Externalities of Carbon Dioxide, Center for the Study of Carbon Dioxide and Global Change.* http://www.coordinationrurale.fr/images/sections/13-10-21_Craig_Idso-The_positive_externalities_of_carbon_dioxide_monetary_ben efits_on_food_production.pdf.

Institute for Energy Research. 2016. "EPA Mandates Renewable Fuel Levels Above 10 Percent Blend Wall."

https://instituteforenergyresearch.org/analysis/epa-mandates-renewable-fuel-levels-10-percent-blend-wall/.

Investopedia. 2016. "What is Purchasing Power Parity?" *Investopedia*, November 2. https://www.investopedia.com/updates/purchasing-power-parity-ppp/.

—. 2015. "What percentage of the global economy is comprised of the oil & gas drilling sector?" *Investopedia*. https://www.investopedia.com/ask/answers/030915/what-percentage-global-economy-comprised-oil-gas-drilling-sector.asp.

IPCC. 2013. In *Climate Change 2013: The Physical Science Basis. Contribution of Working Group I to the Fifth Assessment Report of the Intergovernmental Panel on Climate Change*, by T. Stocker, D. Qin, G.-K. Plattner, M. Tignor, S.K. Allen, J. Boschung, A. Nauels, Y. Xia, V. Bex and P.M. Midgley. Cambridge: Cambridge University Press. https://www.ipcc.ch/pdf/assessment-report/ar5/wg1/WG1AR5_SPM_FINAL.pdf.

IPCC. 2014b. "Climate Change 2014: Impacts, Adaptation, and Vulnerability. Part A: Global and Sectoral Aspects. Contribution of Working Group II to the Fifth Assessment Report of the Intergovernmental Panel on Climate Change." In *Climate Change 2014*, by C.B. Field, V.R. Barros, D.J. Dokken, K.J. Mach, M.D. Mastrandrea, T.E. Bilir, M. Chatterjee, et al. Cambridge University Press. www.ipcc.ch/pdf/assessment-report/ar5/wg2/WGIIAR5-PartA_FINAL.pdf.

—. 2014. *Fifth Assessment Report (AR5)*. 4 vols. Cambridge: Cambridge University Press. https://www.ipcc.ch/report/ar5/.

—. 2001. *TAR, Working Group II: Chapter 15: Impacts, Adaptation and Vulnerability, 15.2.3.1.1. Change in land use.* http://www.ipcc.ch/ipccreports/tar/wg2/index.php?idp=563 .

JAMSTEC. 2016. "Japan Argo Delayed-mode Data Base MOAA data." http://www.jamstec.go.jp/ARGO/argo_web/argo/?page_id =83&lang=en.

—. 2016a. "Mixed Layer data set of Argo, Grid Point Data." www.jamstec.go.jp/ARGO/argo_web/ancient/MILAGPV/i ndex_e.html.

Jerejeva, S., A. Grinsted, and J. Moore. 2009. "Anthropogenic forcing dominates sea level rise since 1850." *Geophysical Research Letters.* http://onlinelibrary.wiley.com/doi/10.1029/2009GL040216/ full.

Joyce, Mary. 2014. "Biofuels production drives growth in overall biomass energy use over past decade." *Today in Energy.* https://www.eia.gov/todayinenergy/detail.php?id=15451.

Kavuncu, Y.O., and S. D. Knabb. 2005. "Stabilizing greenhouse gas emissions: Assessing the intergeneratgional costs and benefits of the Kyoto Protocol." *Energy Economics* 27: 369-386. http://opac.vimaru.edu.vn/edata/E-Journal/2005/Energy%20economics/v27su3.2.pdf.

Koch, J., J.J Clague, and G. Osborn. 2014. "Alpine glaciers and permanent ice and snow patches in western Canada approach their smallest sizes since the mid-Holocene, consistent with global trends." *The Holocene.* http://journals.sagepub.com/doi/abs/10.1177/09596836145 51214.

Kolbert, Elizabeth. 2014. *The Sixth Extinction.* Henry Holt. https://www.amazon.com/Sixth-Extinction-Unnatural-History/dp/0805092994/ref=cm_cr_arp_d_product_top?ie= UTF8.

Kubota, Y., Katsunori Kimoto, R. Tada, Hirokuni Oda, and Y. Yokoyama. 2010. "Variations of East Asian summer monsoon since the last deglaciation based on Mg/Ca and oxygen isotope of planktic foraminifera in the northern East China Sea." *Paleoceanography and Paleoclimatology* 25 (4). http://onlinelibrary.wiley.com/doi/10.1029/2009PA001891/f ull.

Kullman, Leif. 2000. "20th Century Climate Warming and Tree-limit Rise in the Southern Scandes of Sweden." *AMBIO: A Journal of the Human Environment.* http://www.bioone.org/doi/abs/10.1579/0044-7447-30.2.72.

Kundzewicz, Zbigniew W., Dariusz Graczyk, Thomas Maurer, Iwona Pińskwar, Maciej Radziejewski, Cecilia Svensson, and Małgorzata Szwed. 2009. "Trend detection in river flow series: 1. Annual maximum flow." *Hydrological Sciences Journal.* http://www.tandfonline.com/doi/pdf/10.1623/hysj.2005.50. 5.797.

Lamy, F., C. Ruhlemann, D. Hebbein, and G. Wefer. 2002. "High- and low-latitude climate control on the position of the southern Peru-Chile Current during the Holocene." *Paleoceanography and Paleoclimatology.* http://onlinelibrary.wiley.com/doi/10.1029/2001PA000727/full.

Landrigan, et al. 2017. "Lancet Commission on pollution and health." *The Lancet.* doi:10.1016/S0140-6736(17)32345-0.

Landsea, Chris. 2007. *How might global warming change hurricane intensity, frequency, and rainfall ?* NOAA Hurricane Research Division. http://www.aoml.noaa.gov/hrd/tcfaq/G3.html.

Lawler, Andrew. 2015. "Did Egypt's Old Kingdom Die—or Simply Fade Away?" *National Geographic,* December 24. https://news.nationalgeographic.com/2015/12/151224-egypt-climate-change-old-kingdom-archaeology/.

Li, Peng, ChanghuiPeng, Meng Wang, Weizhong Li, Pengxiang Zhao, Kefeng Wang, Yanzheng Yang, and Qiuan Zhu. 2017. "Quantification of the response of global terrestrial net primary production to multifactor global change." *Ecological Indicators* 245-255. https://www.sciencedirect.com/science/article/pii/S1470160 X17300274.

Li, Y, T. Tornqvist, J. Nevitt, and B. Kohl. 2011. "Synchronizing a sea-level jump, final Lake Agassiz drainage, and abrupt cooling." *Earth and Planetary Science Letters* 41-50. https://www.researchgate.net/profile/Torbjoern_Toernqvist /publication/238504773_Synchronizing_a_sea-level_jump_final_Lake_Agassiz_drainage_and_abrupt_cooling

_8200_years_ago/links/5a006a65a6fdcc82a30d0298/Synchro
nizing-a-sea-level-jump-final-Lake-Agassiz-d.

Liu, Zhengyu, Jiang Zhu, Yair Rosenthal, Xu Zhang, Bette L. Otto-Bliesner, Axel Timmermann, Robin S. Smith, Gerrit Lohmann, Weipeng Zheng, and Oliver Elison Timm. 2014. "The Holocene temperature conundrum." *PNAS* 111. http://www.pnas.org/content/111/34/E3501.short.

Lobell, David B., Wolfram Schlenker, and Justin Costa-Roberts. 2011. "Climate Trends and Global Crop Production Since 1980." *Science* 333: 616-620. http://www2.gi.alaska.edu/~bhatt/Teaching/ATM694.fall20 11/post/Papers/Lobell%20et%20al%20SCIENE%202011%2 0climate%20and%20crop%20production.pdf.

Lomborg, B. 2001. *The Skeptical Environmentalist: Measuring the Real State of the World.* Cambridge University Press. doi:ISBN -13: 978-0521010689.

Lomborg, Bjorn. 2007. *Cool It.* Vintage Books. http://www.lomborg.com/cool-it.

M.MacDonald, Glen, Andrei A.Velichko, Constantine V.Kremenetski, Olga K.Borisov, Aleksandra A.Goleva, Andrei A.Andreev, Les C.Cwynar, et al. 2000. "Holocene Treeline History and Climate Change Across Northern Eurasia." *Quaternary Research* 53 (3). https://www.sciencedirect.com/science/article/pii/S0033589 499921233.

Malthus, Thomas. 1798. *An Essay on the Principle of Population*. London: J. Johnson. http://www.esp.org/books/malthus/population/malthus.pdf.

Manderscheid, R., and H.J. Weigel. 2007. "Drought stress effects on wheat are mitigated by atmospheric CO2 enrichment." *Agronomy for Sustainable Development* 27: 79-87. http://www.co2science.org/articles/V10/N41/B1.php.

Mann, Michael E., Raymond S. Bradley, and Malcolm K. Hughes. 1998. "Global-scale temperature patterns and climate forcing over the past six centuries." *Nature* 392: 779-787. https://www.nature.com/articles/33859.

Maraco, J.P., G.E. Edwards, and M.S.B. Ku. 1999. "Photosynthetic acclimation of maize to growth under elevated levels of carbon dioxide." *Planta* 210: 115-125. http://co2science.org/articles/V3/N11/B2.php.

Marcott, S., J. Shakun, P. Clark, and A. Mix. 2013c. "Supplementary Materials for A Reconstruction of Regional and Global Temperature for the Past 11,300 Years." *Science*. http://science.sciencemag.org/content/suppl/2013/03/07/339.6124.1198.DC1.

Marcott, Shaun A., Jeremy D. Shakun, Peter U. Clark, and Alan C. Mix. 2013. "A Reconstruction of Regional and Global Temperature for the Past 11,300 Years." 1198-1201. https://www2.bc.edu/jeremy-shakun/Marcott%20et%20al.,%202013,%20Science.pdf.

Marcott, Shaun A., Jeremy D. Shakun, Peter U. Clark, and and Alan C. Mix. 2013b. "Summary and FAQ's related to the study by

Marcott et al. (2013, Science)." March 31. http://www.realclimate.org/index.php/archives/2013/03/response-by-marcott-et-al/.

May, Andy. 2017e. "A Holocene Temperature Reconstruction Part 1: the Antarctic." https://andymaypetrophysicist.com/2017/05/31/a-holocene-temperature-reconstruction-part-1-the-antarctic/.

—. 2017f. "A Holocene Temperature Reconstruction Part 2: More Reconstructions." June 6. https://andymaypetrophysicist.com/2017/06/06/a-holocene-temperature-reconstruction-part-2-more-reconstructions/.

—. 2017g. "A Holocene Temperature Reconstruction Part 3: The NH and Arctic." June 8. https://andymaypetrophysicist.com/2017/06/08/a-holocene-temperature-reconstruction-part-3-the-nh-and-arctic/.

—. 2017h. "A Holocene Temperature Reconstruction Part 4: The global reconstruction." June 9. https://andymaypetrophysicist.com/2017/06/09/a-holocene-temperature-reconstruction-part-4-the-global-reconstruction/.

—. 2016. "Can Earth be compared to Venus." https://andymaypetrophysicist.com/can-earth-become-venus/.

—. 2016f. "Climate and Civilization for the past 4,000 years." https://andymaypetrophysicist.com/climate-and-civilization-for-the-past-4000-years/.

—. 2015b. "Climate and Human Civilization over the last 18,000 years, updated." November 26. https://andymaypetrophysicist.com/climate-and-human-civilization-over-the-last-18000-years/.

—. 2016d. "CO2, Good or Bad." https://andymaypetrophysicist.com/co2-good-or-bad/.

May, Andy. 2016h. "Comparing early 20th Century warming to late 20th Century warming." https://wordpress.com/posts/andymaypetrophysicist.com.

May, Andy. 2016e. "Detection and Attribution of Man-Made Climate Change." https://andymaypetrophysicist.com/detection-and-attribution-of-man-made-climate-change/.

—. 2017b. "Energy and Society from now until 2040." link: https://andymaypetrophysicist.com/2017/01/05/energy-and-society-from-now-until-2040/.

May, Andy. 2017c. "Highlights of the 2017 Heartland Energy Conference." https://andymaypetrophysicist.com/2017/11/11/highlights-of-the-2017-heartland-energy-conference/.

—. 2015d. "Holocene Thermal Optimum." https://andymaypetrophysicist.com/holocene-thermal-optimum/.

—. 2015c. "New Book: A Disgrace to the Profession, by Mark Steyn." August 28. https://andymaypetrophysicist.com/a-disgrace-to-the-profession/.

May, Andy. 2016c. "Ocean Cycles, The Pause and Global Warming." https://andymaypetrophysicist.com/ocean-cycles-the-pause-and-global-warming/.

—. 2017i. "Oil - Will we run out?" February 17. https://andymaypetrophysicist.com/oil-will-we-run-out/.

May, Andy. 2017d. "Proxy bibliography Excel spreadsheet." https://andymaypetrophysicist.files.wordpress.com/2017/06/reconstruction_references.xlsx.

May, Andy. 2016g. "The Bray (Hallstatt) Cycle." https://wordpress.com/posts/andymaypetrophysicist.com.

—. 2017. "The Greenhouse Effect." https://andymaypetrophysicist.com/the-greenhouse-effect/.

—. 2015. "The Value of Petroleum Fuels." https://andymaypetrophysicist.com/the-value-of-petroleum-fuels/.

—. 2017j. "University of Chicago Air Quality-Life Index." https://andymaypetrophysicist.com/2017/10/07/university-of-chicago-air-quality-life-index/.

—. 2016a. "Using R to Examine Ocean Temperatures." https://andymaypetrophysicist.com/using-r-to-examine-ocean-temperatures/.

Mayewski, Paul A., Eelco Rohling, J. Stager, W. Karlen, Kirk Maasch, D. Meeker, Erick Meyerson, et al. 2004. "Holocene Climate

Variablity." *Quaternary Research* 62 (3): 243-255. https://www.sciencedirect.com/science/article/pii/S0033589 404000870.

Mayewski, Paul A., Loren Meeker, Mark Twickler, Sallie Whitlow, Qinzhao Yang, W. Lyons, and M. Prentice. 1997. "Major features and forcing of high-latitude northern hemisphere atmospheric circulation using a 110,000-year-long glaciochemical series." *Journal of Geophysical Research Oceans* 102 (C12): 26345-26366. http://onlinelibrary.wiley.com/doi/10.1029/96JC03365/full.

Mayr, Ernst. 1974. *Populations, Species and Evolution.* Cambridge: Harvard University Press. https://books.google.com/books?hl=en&lr=&id=STEXkuje BnsC&oi=fnd&pg=PA1&dq=Animal+species+and+evolutio n&ots=YvUi1J072p&sig=kx9nT8evC6Y7GzGB6jrilxpeU- Y#v=onepage&q=Animal%20species%20and%20evolution& f=false.

McIntyre, S. 2013. "Marcott Mystery #1." *Climate Audit*, March 13. https://climateaudit.org/2013/03/13/marcott-mystery-1/.

—. 2013b. "How Marcottian Upticks Arise." March 15. https://climateaudit.org/2013/03/15/how-marcottian- upticks-arise/.

—. 2013c. "The Marcott-Shakun Dating Service." March 16. https://climateaudit.org/2013/03/16/the-marcott-shakun- dating-service/.

McIntyre, Stephen, and Ross McKitrick. 2005. "Hockey sticks, principal components, and spurious significance."

GEOPHYSICAL RESEARCH LETTERS 32. http://www.climateaudit.info/pdf/mcintyre.mckitrick.2005.grl.pdf.

McKitrick, Ross. 2018. "Statement of Ross McKitrick." https://www.rossmckitrick.com/uploads/4/8/0/8/4808045/nyc_lawsuit0.pdf.

Meehl, Gerald A., Aixue Hu, Benjamin D. Santer, and Shang-Ping Xie. 2016. "Contribution of the Interdecadal Pacific Oscillation to twentieth-century global surface temperature trends." *Nature Climate Change* 1005-1008. http://www.nature.com/articles/nclimate3107?cookies=accepted.

Met Office Hadley Centre and Climatic Research Unit. 2017. "Global Average Temperature Series." https://www.metoffice.gov.uk/hadobs/monitoring/index.html.

Met Office Hadley Centre. 2017. "HadCRUT 4 Data: download." https://www.metoffice.gov.uk/hadobs/hadcrut4/data/current/download.html.

Michalek, M. J. 2017. "Examining the progression and termination of Lake Agassiz." https://msu.edu/~michal76/research/407_Geomorphology_Lake%20Agassiz2.pdf.

Miskolczi, Ferenc. 2014. "The Greenhouse Effect and the Infrared Radiative Structure of the Earth's Atmosphere." *Development in*

Earth   *Science.*
http://www.seipub.org/des/paperInfo.aspx?ID=21810.

Mora, C., D. Tittensor, S. Adl, A. Simpson, and B. Worm. 2011. "How
    Many Species Are There on Earth and in the Ocean?" *PLOS
    Biology.*
    http://journals.plos.org/plosbiology/article?id=10.1371/jour
    nal.pbio.1001127.

Morice, Colin. 2017. "MET Office Hadley Centre observations
    datasets."
    https://www.metoffice.gov.uk/hadobs/hadcrut4/data/curren
    t/download.html.

Morner, Nils-Axel. 2012. "Sea level is not rising." SPPI.
    http://scienceandpublicpolicy.org/wp-
    content/uploads/2012/12/sea_level_not_rising.pdf.

Muller, Richard. 2004. "Global Warming Bombshell." *MIT Technology
    Review,*   October   15.
    https://www.technologyreview.com/s/403256/global-
    warming-bombshell/.

Murray, J. L., and A. D. Lopez. 1997. "Mortality by cause for eight
    regions of the world: Global Burden of Disease Study." *The
    Lancet*   349   (1):   1269-1276.
    http://home.cc.umanitoba.ca/~chaser/readings/Global%20
    Burden%20of%20Disease/GBDS%201%20Mortality%20by
    %20cause.pdf.

Murray, J. L., and A. D. Lopez. 1997-2. "Regional patterns of disability-
    free life expectancy and disability-adjusted life expectancy:

Global Burden of Disease Study." *The Lancet* 349 (1): 1347-1352.

NASA Goddard Institute for Space Studies. 2017. "GISS Surface Temperature Analysis (GISTEMP)." http://data.giss.nasa.gov/gistemp/.

NASA. 2017b. "GISS Surface Temperature Analysis." *Goddard Institute for Space Studies.* https://data.giss.nasa.gov/gistemp/maps/.

—. 2017. "SEDAC: Socioeconomic Data and Applications Center." http://sedac.ciesin.columbia.edu/data/set/epi-environmental-performance-index-2016/data-download.

—. 2013. "Venus, In Depth." https://solarsystem.nasa.gov/planets/venus/indepth.

National Snow & Ice Data Center. 2017. "SOTC: Sea Ice." http://nsidc.org/cryosphere/sotc/sea_ice.html.

National Snow and Ice Data Center. 2017. "About the Cryosphere." http://nsidc.org/cryosphere/.

NCDC/NOAA. 2017b. "Greenland Ice Core Data." ftp://ftp.ncdc.noaa.gov/pub/data/paleo/icecore/greenland/.

Newell, R.E., and T.G Dopplick. 1979. "Questions Concerning the Possible Influence of Anthropogenic CO2 on Atmospheric Temperature." *J. Applied Meterology* 18: 822-825. http://journals.ametsoc.org/doi/pdf/10.1175/1520-0450(1979)018%3C0822%3AQCTPIO%3E2.0.CO%3B2.

Nielsen, S., N. Koc, and X. Crosta. 2004. "Holocene climate in the Atlantic sector of the Southern Ocean: Controlled by insolation or oceanic circulation?" *Deology* 32: 317-320. https://www.researchgate.net/profile/Xavier_Crosta/publication/240669732_Holocene_climate_in_the_Atlantic_sector_of_the_Southern_Ocean_Controlled_by_insolation_or_oceanic_circulation/links/00b4951cc0b29133a7000000.pdf.

Niven, Robert K. 2005. "Ethanol in gasoline: environmental impacts and sustainablity review article." *Renewable and Sustainable Energy Reviews* 9 (6): 535-555. https://www.sciencedirect.com/science/article/pii/S1364032104000784.

NOAA. 2017. "Global Tropical Moored Buoy Array." https://www.pmel.noaa.gov/tao/drupal/disdel/.

NOAA. 2017c. *NCEP/NCAR Reanalysis 1: Summary.* NOAA Earth System Research Laboratory. https://www.esrl.noaa.gov/psd/data/gridded/data.ncep.reanalysis.html.

NOAA. 2017d. *Storm Events Database Version 3.0.* NOAA. https://www.ncdc.noaa.gov/news/storm-events-database-version-30.

NOAA/NESDIS/STAR. 2017. "Laboratory for Satellite Altimetry/Sea Level Rise." https://www.star.nesdis.noaa.gov/sod/lsa/SeaLevelRise/.

North, Gerald R. 2006. *Surface Temperature Reconstructions for the Last 2,000 Years.* Testimony to the House of Representatives Energy and Commerce Committee, U.S. House of Representatives.

http://www.nationalacademies.org/images/testimony7-19-06.pdf.

Oldenborgh, G., Karin van der Wiel, Antonia Sebastian, Roop Singh, Julie Arrighi, Friederike Otto, Karsten Haustein, Sihan Li, Gabriel Vecchi, and Heidi Cullen. 2017. "Attribution of extreme rainfall from Hurricane Harvey, August 2017." *Environmental Research Letters.* http://iopscience.iop.org/article/10.1088/1748-9326/aa9ef2.

Ottman, M.J., B.A. Kimball, P.J. Pinter Jr., G.W. Wall, R.L. Vanderlip, S.W. Leavitt, R.L. LaMorte, A.D. Matthias, and T.J. Brooks. 2001. "Elevated CO2 increases sorghum biomass under drought condition." *New Phytologist* 150: 261-273. http://co2science.org/articles/V4/N20/B1.php.

Pachauri, Shonali, and Leiwen Jiang. 2008. "The Household Energy Transition in India and China." *Energy Policy.* http://pure.iiasa.ac.at/8772/1/IR-08-009.pdf.

Pahnke, K., and J.P. Sachs. 2006. "Sea surface temperatures of southern midlatitudes 0–160 kyr B.P." *Paleooceanography and Paleoclimatology* 21 (2). http://onlinelibrary.wiley.com/doi/10.1029/2005PA001191/full.

Pelejero, C., and J. Grimalt. 1999. "High-resolution UK37 temperature reconstructions in the South China Sea over the past 220 kyr." *Paleoceanography* 14 (2). http://onlinelibrary.wiley.com/doi/10.1029/1998PA900015/pdf.

Petit, J.R., J. Jouzel, D. Raynaud, J. Barkov, M. Barnola, I.Basile, M. Bender, and J. Chappellaz. 1999. "Climate and atmospheric history of the past 420,000 years from the Vostok ice core, Antarctica." *Nature* 399: 429-436. https://cloudfront.escholarship.org/dist/prd/content/qt7rx4 413n/qt7rx4413n.pdf.

Pielke Jr., R. 2010. *The Climate Fix, What Scientists and Politicians won't tell you about global warming.* New York, New York: Basic Books. link: http://sciencepolicy.colorado.edu/publications/special/climat e_fix/index.html.

Pielke Jr., Roger. 2017. "STATEMENT OF DR. ROGER PIELKE, JR. to the COMMITTEE ON SCIENCE, SPACE, AND TECHNOLOGY of the UNITED STATES HOUSE OF REPRESENTATIVES." U.S. House of Representatives, Washington, DC. https://science.house.gov/sites/republicans.science.house.go v/files/documents/HHRG-115-SY-WState-RPielke-20170329.pdf.

Pimentel, D., M. Tort, L. D'Anna, A. Krawic, J. Berger, J. Rossman, F. Mugo, et al. 1998. "Ecology of Increasing Disease." *BioScience* 48 (10): 817-826. http://www.jstor.org/stable/1313393?seq=1#page_scan_tab _contents.

Pimentel, D., S. Cooperstein, H. Randell, Filiberto, D., S. Sorrentio, B. Kaye, et al. 2007. "Ecology of Increasing Diseases: Population Growth and Environmental Degradation." *Human Ecology* 35: 653-668. http://erepository.uonbi.ac.ke/bitstream/handle/11295/5508 8/Pimental_Ecology%20of%20increasing%20disease,%20pop ulation%20growth,%20and%20environmental%20degradation?sequence=1&isAllowed=y.

Pisaric, M., C. Holt, and J. Szeicz. 2003. "Holocene treeline dynamics in the mountains of northeastern British Columbia, Canada, inferred from fossil pollen and stomata." *The Holocene.* http://journals.sagepub.com/doi/abs/10.1191/0959683603hl599rp.

PK. 2017. "S&P 500 Return Calculator, with Dividend Reinvestment." *DQYDJ*, December 19. https://dqydj.com/sp-500-return-calculator/.

Popovich, Nadja. 2017. "Today's Energy Jobs Are in Solar, Not Coal." *New York Times.* https://www.nytimes.com/interactive/2017/04/25/climate/todays-energy-jobs-are-in-solar-not-coal.html?smid=tw-nytimesbusiness&smtyp=cur.

Porter, J.R., L. Xie, A.J. Challinor, K. Cochrane, S.M. Howden, M.M. Iqbal, D.B. Lobell, and M.I. Travasso. 2014. "2014: Food security and food production systems." In *Climate Change 2014: Impacts, Adaptation, and Vulnerability. Part A: Global and Sectoral Aspects. Contribution of Working Group II to the Fifth Assessment Report of the Intergovernmental Panel on Climate Change,* by C.B. Field, V.R. Barros, D.J. Dokken, K.J. Mach, M.D. Mastrandrea, T.E. Bilir, M. Chatterjee, et al., 485-533. Cambridge University Press. http://www.ipcc.ch/pdf/assessment-report/ar5/wg2/WGIIAR5-Chap7_FINAL.pdf.

Pritchett, Lant, and Lawrence Summers. 1996. "Wealthier is Healthier." *Journal of Human Resources.* http://www.jstor.org/stable/146149?seq=1#page_scan_tab_contents.

Rahmstorf, S., J. Box, G. Feulner, M. Mann, A. Robinson, S. Rutherford, and E. Schaffernicht. 2015. "Exceptional twentieth-century slowdown in Atlantic Ocean overturning circulation." *Nature Climate Change.* http://www.nature.com/articles/nclimate2554.

Ramsayer, Kate. 2014. "Antarctic Sea Ice Reaches New Record Maximum." *NASA Ice,* October 7. https://www.nasa.gov/content/goddard/antarctic-sea-ice-reaches-new-record-maximum.

Ray, Richard, and Bruce Douglas. 2011. "Experiments in reconstructing twentieth-century sea levels." *Progress in Oceanography* 91 (4): 496-515. https://www.sciencedirect.com/science/article/pii/S0079661111000759.

Reimer, P.J., M G L Baillie, E Bard, A Bayliss, J W Beck, P G Blackwell, C Bronk Ramsey, et al. 2009. "IntCal09 and Marine09 Radiocarbon Age Calibration Curves, 0-50,000 Years cal BP." In *Radiocarbon,* by Edouard Bard. University of Arizona. https://journals.uair.arizona.edu/index.php/radiocarbon/issue/view/181.

Reiny, Samson. 2016. "CO2 is making the Earth greener - for now." *NASA Global Climate Change.* https://climate.nasa.gov/news/2436/co2-is-making-earth-greenerfor-now/.

Rice, Tim. 2011. "Biofuels are driving food prices higher." *The Guardian*. https://www.theguardian.com/global-development/poverty-matters/2011/jun/01/biofuels-driving-food-prices-higher.

Ridley, Matt. 2010. *The Rational Optimist: How Prosperity Evolves*. HarperCollins. http://www.rationaloptimist.com/.

Rignot, Eric, and Pannir Kanagaratnam. 2006. "Changes in the Velocity Structure of the Greenland Ice Sheet." *Science* 311: 986-990. https://cloudfront.escholarship.org/dist/prd/content/qt7f48j474/qt7f48j474.pdf.

Risser, M, and M Wehner. 2017. "Attributable Human-Induced Changes in the Likelihood and Magnitude of the Observed Extreme Precipitation during Hurricane Harvey." *Geophysical Research Letters*. http://onlinelibrary.wiley.com/doi/10.1002/2017GL075888/full.

Ritchie, Justin, and Hadi Dowlatabadi. 2017. "Why do climate change scenarios return to coal?" *Energy* 140: 1276-1291. https://www.sciencedirect.com/science/article/pii/S0360544217314597#!

Robredo, A., Perez-Lopez, Sainz de le Maza U., Gonzalez-Moro, B. H., M., Mena-Petite, A. Lacuesta, and A. Munoz-Rueda. 2007. "Elevated CO2 alleviates the impact of drought on barley improving water status by lowering stomatal conductance and delaying its effects on photosynthesis." *Environmental and Experimental Botany* 59: 252-263. http://www.co2science.org/articles/V10/N18/B2.php.

Rosenthal, Yair, Braddock Linsley, and Delia Oppo. 2013. "Pacific Ocean Heat Content During the Past 10,000 years." *Science*. http://science.sciencemag.org/content/342/6158/617.

Roser, Max. 2017. "Life Expectancy." *Our World in Data*. https://ourworldindata.org/life-expectancy.

Roser, Max, and Hannah Ritchie. 2018b. "Food per person." *Our World in Data*. https://ourworldindata.org/food-per-person.

—. 2018c. "Hunger and Undernourishment." *Our World in Data*. https://ourworldindata.org/hunger-and-undernourishment.

—. 2018. "Yields and Land Use in Agriculture." *Our World in Data*. https://ourworldindata.org/yields-and-land-use-in-agriculture#yields-across-the-world-in-key-crops.

Ryan, William, and Walter Pittman. 2000. *Noah's Flood: The New Scientific Discoveries About The Event That Changed History*. Simon and Schuster. https://www.amazon.com/Noahs-Flood-Scientific-Discoveries-Changed/dp/0684859203/ref=sr_1_1?ie=UTF8&qid=14468 44085&sr=8-1&keywords=Noah%27s+Flood.

Sachs, J. P. 2007. "Cooling of Northwest Atlantic slope waters during the Holocene." *Geophysical Reasearch Letters* 34 (3). http://onlinelibrary.wiley.com/doi/10.1029/2006GL028495/full.

Sarnthein, M., R. Gersonde, S. Pflaumann, U. Niebler, R. Spielhagan, J. Thiede, G. Wefer, and M. Weinelt. 2003. "Overview of Glacial Atlantic Ocean Mapping (GLAMAP 2000)." *Paleooceanography and Paleoclimatology* 18 (2).

http://onlinelibrary.wiley.com/doi/10.1029/2002PA000769/full.

Schliebe, S., Ø. Wiig, A. Derocher, and N Lunn. 2008. *Ursus maritimus, Polar Bear.* IUCN "Red" list, IUCN. https://polarbearscience.files.wordpress.com/2016/09/ursus-maritimus_iucn-red-list-2008_t22823a9391171-en_red-list-generated.pdf.

Schmidt, G.A. 2017a. "Datasets and Images." *NASA.* https://data.giss.nasa.gov/.

Schouten, E.C. Hopmans, E. Schefu, and Damst Sinninghe. 2003. "Distributional variations in marine crenarchaeotal membrane lipids: a new tool for reconstructing ancient sea water temperatures?" *Earth Planet Sci Lett* 204. https://www.researchgate.net/publication/222579577_Schouten_S_Hopmans_EC_Schefu_E_Sinninghe_Damst_JS_Distributional_variations_in_marine_crenarchaeotal_membrane_lipids_a_new_tool_for_reconstructing_ancient_sea_water_temperatures_Earth_Planet_Sci_Lett_.

Schrover, Mariou. 2008. "History of International Migration." Universiteit Leiden. http://www.let.leidenuniv.nl/history/migration/chapter131.html.

Schwartz, J.Z. 2011. "Food prices: Eating the Cost of Logistics." http://blogs.worldbank.org/latinamerica/food-prices-eating-the-cost-of-logistics.

Selby, Jan, Omar S. Dahi, Christiane Fröhlich, and Mike Hulme. 2017. "Climate change and the Syrian civil war revisited." *Political Geography* 232-244. https://www.sciencedirect.com/science/article/pii/S0962629 816301822.

Seppa, H., and D. H. Antonsson. 2005. "Low-frequency and high-frequency changes in temperature and effective humidity during the Holocene in south-central Sweden: implications for atmospheric and oceanic forcings of climate." *Climate Dynamics* 25 (2). https://link.springer.com/article/10.1007/s00382-005-0024-5.

Severinghaus, Jeffrey P., Todd Sowers, Edward J. Brook, Richard B. Alley, and Michael L. Bender. 1998. "Timing of abrupt climate change at the end of the Younger Dryas interval from thermally fractionated gases in polar ice." *Nature*, January 8: 141-146. http://shadow.eas.gatech.edu/~jean/paleo/Severinghaus_199 8.pdf.

Smith, D. K., L. Ricclardull, C. Mears, and K. Hilburn. 2017. *A 25-YEAR SATELLITE MICROWAVE MEAN TOTAL PRECIPITABLE WATER DATA SET FOR USE IN CLIMATE STUDY.* RSS. https://www.eumetsat.int/website/wcm/idc/idcplg?IdcServi ce=GET_FILE&dDocName=PDF_CONF_P_S7_08_SMIT H_V&RevisionSelectionMethod=LatestReleased&Rendition= Web.

Stone, Lawrence. 2013. *The Family, Sex and Marriage in England 1500-1800*. ACLS Humanities.

Strauss, G. 2012. "AAA warns E15 gasoline could cause car damage." *USA Today*, November 30. https://www.usatoday.com/story/news/nation/2012/11/30/aaa-e15-gas-harm-cars/1735793/.

Sud, Y. C., G. K. Walker, and K. M. Lau. 1999. "Mechanisms Regulating Sea-Surface Temperatures and Deep Convection in the Tropics." *Geophysical Research Letters* 26 (8): 1019-1022. http://onlinelibrary.wiley.com/doi/10.1029/1999GL900197/full.

Talley, L. D. 1996. "Antarctic Intermediate Water in the South Atlantic." In *The South Atlantic*, 219-238. Springer. https://link.springer.com/chapter/10.1007/978-3-642-80353-6_11.

Thouret, Jean-Claude, Thomas Van der Hammenb, Barry Salomons, and Etienne Juvigné. 1996. "Paleoenvironmental Changes and Glacial Stades of the Last 50,000 Years in the Cordillera Central, Colombia." *Quaternary Research* 1-18. https://www.sciencedirect.com/science/article/pii/S0033589496900393.

Tinner, Willy, Brigitta Ammann, and Peter Germann. 1996. "Treeline Fluctuations Recorded for 12,500 Years by Soil Profiles, Pollen, and Plant Macrofossils in the Central Swiss Alps." *Arctic and Alpine Research* 28 (2). http://www.jstor.org/stable/1551753?seq=1#page_scan_tab_contents.

Tirone, Jonathan. 2013. "Voestalpine Choses Texas for Record $718 Million Pellet Plant." *Bloomberg.*

https://www.bloomberg.com/news/articles/2013-03-
13/voestalpine-choses-texas-for-record-718-million-pellet-
plant.

TPCO. 2015. "Project Introduction."
http://www.tpcoamerica.com/about-us.html.

Tzedakis, P., M. Crucifix, and E. Wolff. 2017. "A simple rule to
determine which insolation cycles lead to interglacials." *Nature*
542: 427-432.
http://eprints.esc.cam.ac.uk/3856/1/nature21364.pdf.

U.S. Census Bureau. 2017. "International Programs (Data)."
https://www.census.gov/data-
tools/demo/idb/informationGateway.php.

U.S. Energy Department. 2017. "U.S. Energy and Employment
Report."
https://energy.gov/sites/prod/files/2017/01/f34/2017%20
US%20Energy%20and%20Jobs%20Report_0.pdf.

UNICEF. 2017b. "Global Overview Child Malnutrition."
https://public.tableau.com/profile/unicefdata#!/vizhome/Joi
ntMalnutritionEstimates2017Edition-Wide/Base.

UNICEF, and WHO. 2017. "Progress on sanitation and Drinking
Water."
http://files.unicef.org/publications/files/Progress_on_Sanitat
ion_and_Drinking_Water_2015_Update_.pdf.

United Nations Climate Change. 2013. "Kyoto Protocol."
http://unfccc.int/kyoto_protocol/items/2830.php.

—. 2018. "The Paris Agreement." http://unfccc.int/paris_agreement/items/9485.php.

United Nations Population Division. 2017. "World Population Prospects 2017." https://esa.un.org/unpd/wpp/.

University Of Illinois At Urbana-Champaign. 2004. "Shutdown Of Circulation Pattern Could Be Disastrous, Researchers Say." December 20. https://www.sciencedaily.com/releases/2004/12/041219153 611.htm.

Upsalla Conflict database. 2017. http://ucdp.uu.se/downloads/.

USGS. 2014. "Southern Beaufort Sea Polar Bear Population Declined in the 2000s." https://www.usgs.gov/news/southern-beaufort-sea-polar-bear-population-declined-2000s.

Veisz, O., S. Bencze, K. Balla, G. Vida, and Z. Bedo. 2008. "Change in water stress resistance of cereals due to atmospheric CO_2 enrichment." *Cereal Research Communications* 36. doi:10.1556/CRC.36.2008.Suppl.1.

Vinas, Maria-Jose. 2015. "NASA Study: Mass Gains of Antarctic Ice Sheet Greater than Losses." October 30. https://www.nasa.gov/feature/goddard/nasa-study-mass-gains-of-antarctic-ice-sheet-greater-than-losses.

Vinós, Javier. 2017. "Nature Unbound III: Holocene climate variability (Part A)." *Climate Etc.,* April 30.

https://judithcurry.com/2017/04/30/nature-unbound-iii-holocene-climate-variability-part-a/.

—. 2018. "Nature Unbound VIII - Modern Global Warming." *Climate, Etc.* https://judithcurry.com/2018/02/26/nature-unbound-viii-modern-global-warming/.

Vinós, Javier. 2016. "Impact of the ~2400 year solar cycle on climate and human societies." https://judithcurry.com/2016/09/20/impact-of-the-2400-yr-solar-cycle-on-climate-and-human-societies/.

Vinther, B. M., S. L. Buchardt, H. B. Clausen, D. Dahl-Jensen, S. J. Johnsen, D. A. Fisher, R. M. Koerner, et al. 2009. "Holocene thinning of the Greenland ice sheet." *Nature* 461. http://users.clas.ufl.edu/rrusso/gly6932/vinther_etal_nature09.pdf.

Von der Bouchard, H., and B. Surk. 2016. "Russia's boom (farming) economy." *Politico.* https://www.politico.eu/article/russias-boom-farming-economy/.

Walker, MIKE, SIGFUS JOHNSEN, UNE OLANDER RASMUSSEN, TREVOR POPP, JØRGEN-PEDER STEFFENSEN, PHIL GIBBARD, WIM HOEK, et al. 2009. "Formal definition and dating of the GSSP (Global Stratotype Section and Point) for the base of the Holocene using the Greenland NGRIP ice core,and selected auxiliary records." *JOURNAL OF QUATERNARY SCIENCE* 24: 3-17. http://onlinelibrary.wiley.com/doi/10.1002/jqs.1227/full.

Watling, J.R, and M.C. Press. 1997. "How is the relationship between C4 cereal Sorghum bicolor and the C3 root hemi-parasites

Striga hermonthica and Striga asiatica affected by elevated CO2?" *Plant, Cell and Environment* 20: 1292-1300. http://co2science.org/articles/V3/N1/B2.php.

Weißbach, D., G. Ruprecht, A. Huke, K. Czerski, S. Gottlieg, and A. and Hussein. 2013. "Energy intensities, EROIs (energy returned on invested), and energy payback times of electricity generating power plants." *Energy* 210-221. http://homepages.uc.edu/~becktl/shaka-eroi.pdf.

Welitzkin, Paul. 2016. "Massive pipe plant to boost Chinese FDI in Texas." *China Daily.* http://usa.chinadaily.com.cn/us/2016-05/09/content_25144601.htm.

Werne, Josef P., David Hollander, Timothy Lyons, and Larry Peterson. 2000. "Climate-induced variations in productivity and planktonic ecosystem structure from the Younger Dryas to Holocene in the Cariaco Basin, Venezuela." *Paleoceanography and Paleoclimatology.* http://onlinelibrary.wiley.com/doi/10.1029/1998PA000354/full.

Weyant, J.P., and J. Hill. 1999. "The Costs of the Kyoto Protocol: A Multi-Model Evaluation." *The Energy Journal.* https://web.stanford.edu/group/emf-research/docs/emf16/CostKyoto.pdf.

WHO. 2014c. "7 million premature deaths annually linked to air pollution." http://www.who.int/mediacentre/news/releases/2014/air-pollution/en/.

—. 2016. "An estimated 12.6 million deaths each year are attributable to unhealthy environments." http://www.who.int/mediacentre/news/releases/2016/death s-attributable-to-unhealthy-environments/en/.

WHO. 2017c. "Global health observatory data repository." http://apps.who.int/gho/data/view.main.NUTWHOSEVW ASTINGv?lang=en.

—. 2016b. "Health statistics and information, Cause Specific Mortality, Estimates for 2000-2015." http://www.who.int/healthinfo/global_burden_disease/estim ates/en/index1.html.

—. 2017. "Household air pollution and health." http://www.who.int/mediacentre/factsheets/fs292/en/.

—. 2017b. "Tobacco." http://www.who.int/mediacentre/factsheets/fs339/en/.

WHO, UNICEF, World Bank, and UN-DESA Population Division. 2014. "Levels and Trends in child mortality 2014." http://www.who.int/maternal_child_adolescent/documents/l evels_trends_child_mortality_2014/en/.

Williams, D.R. 2016. "Venus Fact Sheet." *NASA*. https://nssdc.gsfc.nasa.gov/planetary/factsheet/venusfact.ht ml.

Williams, Matt. 2015. "What is the Average Surface Temperature on Venus." *Universe Today*. https://www.universetoday.com/14306/temperature-of-venus/.

Willis, Josh. 2010. "Can in situ floats and satellite altimeters detect long-term changes in Atlantic Ocean overturning?" *Geophysical Research Letters* 37 (6). http://onlinelibrary.wiley.com/doi/10.1029/2010GL042372/full.

World Bank. 2017b. "GDP, All Countries and Economies." https://data.worldbank.org/indicator/NY.GDP.MKTP.CD.

—. 2017. "International Comparison Program Database." https://data.worldbank.org/indicator/NY.GDP.PCAP.PP.CD.

World Food Programme. 2017. "10 Facts About Nutrition in China." https://www.wfp.org/stories/10-facts-about-nutrition-china.

Wouters, B., D. Chambers, and E. Schrama. 2008. "GRACE observes small-scale mass loss in Greenland." *Geophysical Research Letters* 35 (20). http://onlinelibrary.wiley.com/doi/10.1029/2008GL034816/full.

Wuebbles, D.J., D.W. Fahey, K.A. Hibbard, D.J. Dokken, B.C. Stewart, and T.K. Maycock. 2017. *Climate Science Special Report.* Washington, DC: U.S. Global Change Research Program. https://science2017.globalchange.gov/.

Wyatt, Marcia. 2017. "Wyatt on Earth." http://www.wyattonearth.net/publicationsstadiumwave.html.

Xu, J., A. Holbourne, W. Kuhnt, Z. Jian, and H. Kawamura. 2008. "Changes in the thermocline structure of the Indonesian outflow during Terminations I and II." *Earth and Planetary Science Letters* 273 (1): 152-162. https://www.sciencedirect.com/science/article/pii/S0012821X08003981.

Zhang, Peng, Olivier Deschenes, Kyle Meng, and Junjie Zhang. 2018. "Temperature effects on productivity and factor reallocation: Evidence from a half million chinese manufacturing plants." *Journal of Environmental Economics and Management* 88: 1-17. https://www.sciencedirect.com/science/article/pii/S0095069617304588.

Zhu, Zaichun, Shilong Piao, Ranga B. Myneni, Mengtian Huang, Zhenzhong Zeng, Josep G. Canadell, Philippe Ciais, et al. 2016. "Greening of the Earth and its drivers." *Nature Climate Change* 6: 791-795. https://www.nature.com/articles/nclimate3004.

Zuckerman, Laura. 2018. "Eastern U.S. cougar declared extinct 80 years after last sighting." *Reuters*, January 23. https://www.reuters.com/article/us-usa-cougars/eastern-u-s-cougar-declared-extinct-80-years-after-last-sighting-idUSKBN1FC06Z.

Zwally, H. Jay, Jun LI, John W. ROBBINS, Jack L. SABA, Donghui YI, and Anita C. BRENNER. 2015. "Mass gains of the Antarctic ice sheet exceed losses." *Journal of Glaciology* 61 (230): 1019-1036. https://www.cambridge.org/core/services/aop-cambridge-core/content/view/983F196E23C3A6E7908E5FB32EB42268/S0022143000200221a.pdf/mass_gains_of_the_antarctic_ice_sheet_exceed_losses.pdf.

Zwally, H., and M. Giovinetto. 2011. "Overview and Assessment of Antarctic Ice-Sheet Mass Balance Estimates: 1992–2009." *Surveys in Geophysics* 32 (4): 351-376. https://link.springer.com/article/10.1007/s10712-011-9123-5.

LIST OF TABLES

Table 1. Worldwide Deaths from the World Health Organization (WHO)..10

Table 2. Primary energy sources for producing electricity in the U.S. in 2016, data source (EIA 2017b)..73

Table 3. Three recent periods of rapid warming in the Vinther reconstruction compared to comparable periods in the Antarctic Vostok reconstruction, data sources (Vinther, et al. 2009) and (Petit, et al. 1999). "Year BP" means years before 1950.......................................94

Table 4. Temperature proxy summary...97

Table 5. Distribution of the temperature proxies........................107

Table 6. World technically recoverable oil and gas. Source (May 2017i). 146

Table 7. U.S. Hurricanes 1850-2017 by category. Data source: NOAA Hurricane Research Division 183

Table 8. Earth's Ice Sheets and the effect of eustatic sea level rise. Sources: NSIDC, NASA, and the IPCC WG1 AR5 Chapter 13 (Table 13.1 and summary) 202

LIST OF FIGURES

Figure 1. Average ocean temperature and average ocean mixed layer temperature, data sources: JAMSTEC (May 2016a) (JAMSTEC 2016) (JAMSTEC 2016a). 5

Figure 2. 2016 GDP/person plotted against the NASA environmental performance index, data sources: Environmental Performance Index: (NASA 2017), GDP in PPP$: (World Bank 2017). 7

Figure 3. World infectious disease death rates, by year, data sources: blue curves (Bulatao 1993) and the black heavy dash curve, (WHO 2016b). 11

Figure 4. Age adjusted U.S. death rate due to cancer, data from (CDC 2017).. 13

Figure 5. World life expectancy at birth by year, data source: (World Bank 2017). 16

Figure 6. World Bank Energy used per person versus World Bank life expectancy, data source: (World Bank 2017). 17

Figure 7. Energy used per person versus GDP per person for 2011. The whole world is compared to affluent countries, data source: (World Bank 2017). 18

Figure 8. Life expectancy is compared to the percent of the life spent disabled for different areas of the world, data source: (Murray and Lopez 1997-2).. 19

Figure 9. Measures of human prosperity and well-being, data source (World Bank 2017)..26

Figure 10. U.N. world population growth, data source: (United Nations Population Division 2017). 33

Figure 11. Birth and death rates for four countries, data source: (World Bank 2017)...34

Figure 12. World population growth as a percent of population and in people added, data source: (U.S. Census Bureau 2017). 35

Figure 13. Life expectancy versus GDP per person, data source (Roser, Life Expectancy 2017) used under the CC BY-SA license. 36

Figure 14. Stunted and underweight children under 5, data source WHO data repository (WHO 2017c). 41

Figure 15. Worldwide conflicts where a state government is one participant, data source (Upsalla Conflict database 2017)......................42

Figure 16. World Agricultural Production by year, data Source (FAO Statistics Division 2017). 43

Figure 17. Least developed country agricultural production by year, data source (FAO Statistics Division 2017)..44

Figure 18. Kilocalories (food Calories) of food per person around the world, data source (FAO Statistics Division 2017). 45

Figure 19. FAO Inflation corrected world food prices (FAO 2017c). 46

Figure 20. U.S. Biofuel growth (Source: (Joyce 2014)), used with permission. 47

Figure 21. Map of annual global warming intensity in °C. The change from the average from 1951-1970 to the average from 2011-2017, source NASA (NASA 2017b). 49

Figure 22. Plot of total wheat yield from 146 countries, data from Ourworldindata.org (Roser and Ritchie 2018) and the FAO (FAO Statistics Division 2017). 50

Figure 23. Plots of studies of the impact of climate change on crop yields, source IPCC AR5 WGII, Chapter 7, page 492 (Porter, et al. 2014). Used with permission. 51

Figure 24. Wheat yields in four countries, data source Ourworldindata.org (Roser and Ritchie 2018) and the FAO (FAO 2017b). 54

Figure 25. IPCC AR5 surface temperature projections (data source: IPCC AR5, data downloaded from the KNMI climate explorer). 62

Figure 26. The effect of the Paris Agreement on U.S. jobs, source Heritage.org (Dayaratna, Loris and Kreutzer 2016). Used with permission. 69

Figure 27. Primary energy (mainly coal, natural gas, petroleum, and renewables) consumed 74

Figure 28. World energy consumption, data source (BP 2017). 75

Figure 29. Global surface temperature records, (Met Office Hadley Centre and Climatic 81

Figure 30. Average ocean temperature and the average ocean mixed layer temperature, data sources: JAMSTEC (May 2016a) (JAMSTEC 2016). 85

Figure 31. Global merged land/sea surface temperatures and global sea surface..87

Figure 32. Vinther Greenland temperature reconstruction, elevation corrected, compared to the Alley GISP2 reconstruction, which is not elevation corrected. 90

Figure 33. Google Earth Pro map of Greenland showing the locations of the ice cores discussed in the text, data from U.S. Department of State Geographer and Landsat/Copernicus Image IBCAO, used with permission..91

Figure 34. Marcott Holocene temperature reconstruction, data source Marcott, et al., 2013 supplementary material. 96

Figure 35. Insolation changes for the past 40,000 years. The background color represents the obliquity changes to the Earth's axis. The curves represent the changes to the total insolation at 60°N and 60°S. The beginning of the Holocene has high polar insolation and high summer insolation. Source: Javier, used with permission. 102

Figure 36. Marcott's final reconstruction (orange-dashed line) compared to our final reconstruction. 104

Figure 37. Plot of the temperature reconstruction of the ODP658C sediment core by deMenocal, et al., 2000 (Science). The core is from just offshore of the African Sahara Desert, as shown in the inset Google Earth Pro map. Map data from US Department of State Geographer, SIO, NOAA, U.S. Navy, NGA and GEBCO. 105

Figure 38. The Antarctic proxies 108

Figure 39. The Antarctic temperature reconstruction. 109

Figure 40. Mid-latitude Southern Hemisphere Proxies 111

Figure 41. The Southern Hemisphere temperature reconstruction. 112

Figure 42. The tropical proxies 113

Figure 43. The tropical temperature reconstruction. 114

Figure 44. The locations of the tropical proxies are indicated by arrows. 115

Figure 45. The Northern Hemisphere proxies. 116

Figure 46. The locations of the Northern Hemisphere proxies . 117

Figure 47. The Northern Hemisphere temperature reconstruction. 118

Figure 48. The Arctic proxies. 119

Figure 49. The Arctic proxy locations 119

Figure 50. The Arctic temperature reconstruction........................ 120

Figure 51. All the regional temperature reconstructions plotted together. The Northern Hemisphere reconstruction is the outlier. . 121

Figure 52. The final area-weighted global reconstruction........... 123

Figure 53. Rates of change in distribution (km/decade) for several marine taxonomic groups for 1900 to 2000. Source IPCC WGII AR5 Technical Summary, page 43, Figure TS.2, used with permission. Note: Positive changes mean the taxa are present over a larger area, generally moving poleward.. 139

Figure 54. The Greenland Ice Sheet Project 2 (GISP2) temperature reconstruction, showing the Younger Dryas (Alley, 2004)................ 143

Figure 55. September Arctic Sea Ice Extent, Data from NSIDC (National Snow & Ice Data Center 2017)... 149

Figure 56. AMOC speed, showing a slight increase from the 1990s to the 2000s. From (Willis 2010) in Geophysical Research Letters, used with permission. 152

Figure 57. The percent of all deaths by month for the U.S., data source ABC News.. 161

Figure 58. Temperature projections to 2100 for two Representative Concentration Pathways (RCP). Source IPCC AR5 The Physical Science Basis, Summary for Policymakers, page 21, SPM.7a, used with permission (IPCC 2013)... 164

Figure 59. HADCRUT4 global temperature reconstruction. Data: HADCRUT 4, (Met Office Hadley Centre and Climatic Research Unit 2017. 166

Figure 60. The EPA heat wave index 1890-2015, source (EPA 2017b). Used with permission. 167

Figure 61. Global Historical Climatology Network stations in the Houston, Texas, area. Source (Risser and Wehner 2017), used with permission. 180

Figure 62. Ten-year averages of U.S. hurricane strength, based on the hurricane central pressure. Data source: NOAA Hurricane Research Division. 182

Figure 63. Global tropical cyclone landfalls through 2016. Source: (R. Pielke Jr. 2017), U.S. House of Representatives testimony, used with permission. 184

Figure 64. Global weather-related disaster losses as a percent of GDP. Source: (R. Pielke Jr. 2017). House of Representatives testimony, used with permission. 185

Figure 65. Sea-level change by (Church and White 2011) in orange, (Jerejeva, Grinsted and Moore 2009) in blue, and (Ray and Douglas 2011) in green. (IPCC 2013) used with permission. 199

Figure 66. The Commonwealth Scientific and Industrial Research Organization (CSIRO) sea level from 1880 to 2015 and the AMO (Atlantic Multidecadal Oscillation) index. Data sources CSIRO and NOAA. 200

Figure 67. Ocean temperature trend from Argo and TRITON data. Data source JAMSTEC MOAA GPV grid. 203

Figure 68. Glacier advance data for the Holocene from (Mayewski, Rohling, et al. 2004). Used with permission. 207

Figure 69. Major climatic events, glacial advances and the Bond hematite-stained grain record for the Holocene. Source (Javier 2016). Used with permission. 208

Figure 70. Church and White Sea Level, 20th century. 212

Figure 71. The Thames Barrier, public domain photo. 213

Figure 72. The Galveston seawall in the early 20th century, an undated public domain image. 214

Figure 73. Galveston seawall under construction in 1905, public domain photo, photographer unknown, uploaded to pinterest (link) by Donna Clayton and Ray Gibson.. 215

Figure 74. The Galveston seawall in 2008, shortly after Hurricane Ike (photo credit: Aurelia May) 216

DISCLAIMER

The material, opinions and positions presented in this book, *Climate Catastrophe! Science or Science Fiction?* are those of Andy May and were developed as a result of significant research on the subject. The publisher, American Freedom Publications LLC, makes no representations or warranties of any kind and assume no liabilities of any kind with respect to the accuracy or completeness of the contents of this book.

The publisher shall not be held liable or responsible to any person or entity with respect to any loss or incidental or consequential damages caused, or alleged to have been caused, directly or indirectly, by the information contained herein.

INDEX

4.2 kyr, 103, 106, 118, 120

5.9 kyr, 103, 104, 112, 117, 120, 124

8.2 kyr, 103, 104, 109, 117, 120, 124, 150

adaptation, 31, 53, 60, 63, 64, 65, 79, 168, 196, 218

Advanced Hydrologic Prediction Service, 179

Agassiz, 91, 92, 133, 153, 155

agricultural yields, 51

AMO, **199**

AMOC, 150, 158

Angela Merkel, 190

Antarctic Polar Front, 110

Antarctic proxies, 108

Antarctic sea ice extent, **197**

Antarctica colder, **197**

Arctic, 39, 92, 94, 103, 118, 120, 123, 149

Argo floats, 85, 204

artificial light, 22

Atlantic meridional overturning circulation, 150, 158

Atlantic Multidecadal Oscillation, 181, 199

Barnosky, 140, 141, 142, 144, 145, 146, 155, 158

biofuels, 47, 50

birth rate, 35

blend wall, 48

Bolivar Peninsula, 216

BP Energy Outlook, 74

Bray cycle, 181, 192, 200, 201, 209

Bronze Age, 106, 124

cancer death rate, 12, 13

CDC, 13, 160, 162, 170

Chile, 110

Church and White sea level, 201

Clausius-Clapeyron, 180, 187, 188, 190

climate models, 59, 67, 76, 78, 126, 176, 178
climate-related deaths, 159
climate-related mortality, 159, 170
cold-related deaths, 15, 160, 163
corn ethanol, 47
Craig Idso, 40, 53
crop yields, 31, 42, 46, 55, 56, 78
Crystal Beach, 216
CSIRO, **201**
Daniel Botkin, 138
David Pimentel, 36
death rate, 35
disabilities, 19
discount rate, 59, 64, 65
disease is increasing, 11
Dixon, et al., 160, 162, 170
Doug Erwin, 137
Dr. Judith Curry, xv
economic models, 60, 78
Egyptian Old Kingdom, 106, 124
EIA, 47, 70, 71
El Niño, 81, 179
Elizabeth Kolbert, 133
ENSO, 179, 180, 181, 187

environmental pollution, 10, 16, 29
Epstein, Alex, 21, 29
equilibrium vapor pressure, 187
European Center for Medium Range Weather Forecasts, 178
eustatic, 198, 199, 201, 202, 209, 218
eustatic sea level, **202**
extinction rate, 133, 136, 138, 141
extreme weather, xiv, 168, 173, 174, 177, 186, 192, 193
FAO, 31, 37, 40, 41, 44, 51, 52, 54, 56
flooding, 171, 188, 189, 193, 195, 216
Florence, 189
Food prices, 48
fossil fuel, xv, 2, 5, 29, 64, 66, 67, 70, 76, 79, 174
Fossil fuels, 8, 73
Friends of the Earth, 186
Gaia, **6**
Galveston, 213, 214, 216, 217
genus, **134**
GISTEMP, **3**
Glaciers, 195, 206, 209, 210, 217

global average temperature, 5, 64, 86, 163

great extinction, xiv, 133, 139, 140, 141, 155

Greek Dark Ages, 103, 112, 118, 120, 128

Greenland ice core, 83, 90, 94, 102, 142, 143

Greenpeace, 138, 186

Gulf Stream, xiv, 133, 153, 154, 155

HADCRUT, **3**

heat-related deaths, 15, 63, 159, 160, 162, 163

Heritage Foundation, 69

Hockey Stick, 81

Holocene, 66, 81, 82, 83, 90, 92, 93, 94, 97, 98, 99, 100, 101, 103, 105, 108, 110, 112, 113, 116, 117, 120, 122, 124, 126, 127, 128, 129, 130, 131, 137, 144, 157, 192, 206, 207, 208, 209, 217

Holocene Climatic Optimum, 83, 92, 98, 99, 103, 106, 112, 113,

117, 120, 122, 124, 127, 128

Houston, 178, 180, 182, 193, 214

Hurricane Carla, 214

Hurricane Harvey, 178, 187

Hurricane Ike, 215, 216

hurricanes, 63, 115, 173, 178, 182, 183, 185, 186, 193, 214

IEA, 70

International Union for the Conservation of Nature, 147

Intertropical Convergence Zone, 105, 122

IPCC, **xv, 1, 15, 31, 46, 48, 51, 52, 53, 54, 55, 56, 60, 61, 62, 63, 64, 66, 78, 139, 154, 159, 163, 165, 167, 168, 173, 178, 190, 195, 198, 199, 202, 204, 211, 212**

iron law of climate policy, 175

Isostatic sea level, **198**

James Hansen, **2**

James Kasting, **2**

JAMSTEC, **4**

Javier, 101, 102, 208

Jeffrey Bromaghin, 148

Jochim Schellenhuber, 190

Kelly Cobb, 23

King Louis XIV, 21, 23

Kip Hansen, 196, 205, 206

Kyoto, 67, 75, 76, 186, 189, 217

Lant Pritchett, 68

Lawrence Stone, 20

Lawrence Summers, 68

Leaf Area Index, 39

life expectancy, 16, 17, 26, 35, 36, 37, 55

Little Ice Age, 66, 77, 83, 93, 98, 99, 106, 117, 120, 121, 123, 124, 127, 128, 137, 176, 181, 188, 192, 200, 201, 208, 209, 210, 217

Lomborg, Bjørn, 6, 9, 11, 12, 13, 19, 20, 25, 33, 34, 37, 38, 50, 138

London, 213

malnourished, 10, 37, 40, 41, 43

Marcott, 83, 84, 92, 93, 95, 96, 97, 98, 99, 103, 105, 106, 107, 121, 125, 126, 127, 128, 129, 208

mass extinction, 134, 136, 137, 138, 139, 141, 155

Matt Ridley, 23, 66

Mayr, **134**

Medieval Warm Period, 123, 128, 206

megafauna extinction, 140, 144, 155

Michael Mann, 81

mitigation, 60, 63, 64, 65, 67, 68, 174

mixed layer, **5**

Monte Carlo, 126

multiplication of labor, 23

NASA, 7, 64, 81, 92, 151, 158, 178, 197, 202, 217

natural disasters, 173, 175, 185

Natural Resources Defense Council, 185

Neoglacial, 83, 103, 109, 113, 114, 120, 122, 209

NERA, 68, 70, 71

net plant productivity, 39

New York Times, **197**

New Zealand, 110

Niño3.4, 180

NOAA, 105, 179, 183, 188

North Sea, 193, 213

Obliquity, 100, 101, 106

ocean depth, **4**

Oregon trail, 23

Paris agreement, 68, 72

Paul Ehrlich, 32, 38, 65

Pielke Jr., Roger, 14, 174, 175, 176, 177, 190, 192

polar bears, 147, 148, 149, 155

population growth, xiv, 9, 31, 32, 35, 55, 133, 196

precession, 100, 101, 122

Pre-pottery Neolithic "B", 104

present value, 65, 79

Primary energy, 72

Professor [David] Pimentel, **8**

Professor Nils-Axel Mörner, 204

purchase-power-parity dollar, **7**

Quaternary, 83, 140, 141, 146, 155

R code, 108

RCP8.5, **197**

record cold temperature in Antarctica, **197**

Reducing fossil fuel, 64

Renland, 91, 92

RFF, 68, 70, 71

Richard Alley, 90

Ridley, Matt, 19, 21, 22, 23, 24

Robert M. DeConto, **197**

Roger Pielke Jr., 66, 174, 178, 190, 197

Roman Warm Period, 103, 118, 124, 128

Ross McKitrick, 82

Sahara, 104, 105, 124, 191

Sahel, 191

sea-level rise

IPCC, **195**

Sherwood Idso, 53

Sir David Attenborough, 6, 32

sixth mass extinction, 137

Southern Ocean, 109, 110

species, **134**

Stephen Hawking, **2**

surface temperatures, 81, 85, 86, 89, 92, 98, 128, 187, 199

Susan Crockford, 147, 155

TEX$_{86}$, 98

Texas, 70, 213, 217

Thames Barrier, 212, 213

The Day after Tomorrow, 154

thermal energy, **2**, **3**, **84**, **85**, **100**, **115**, **201**

Thomas Malthus, 32, 65

time value of money, xiii, 64, 65, 76, 218

TRITON, 85

tuberculosis, 12, 20

U.S. National Center for Atmospheric Research, 178

UK Met Office, 81, 86

UK'37, 98

United Kingdom, 189

United States Fish and Wildlife Service, 147

University of East Anglia Climate Research Unit, 81

Upper Paleolithic, 144, 146, 155

Uppsala Conflict database, 42

Venus, **2**

Vinther, et al., 90, 92

Vostok ice core, 94

wheat yields, 51, 54

WHO, 40

World Bank, 7, 17, 26, 47, 61

World food prices, **47**

World GDP, 61

World Health Organization, 9, 10, 12, 14, 170, 172

Younger Dryas, 83, 117, 142, 143

Zaichun Zhu, **39**

Zwally, et al., 217

$\delta^{18}O$, 98, 102